Oxford Applied Mathematics and Computing Science Series

CW00377130

General Editors
J. N. Buxton, R. F. Churchhouse, and A. B. Tayler

OXFORD APPLIED MATHEMATICS AND COMPUTING SCIENCE SERIES

Pete Thomas, Hugh Robinson, and Judy Emms
Open University

Abstract Data Types
Their Specification, Representation, and Use

CLARENDON PRESS · OXFORD

Oxford University Press, Walton Street, Oxford OX2 6DP
Oxford New York Toronto
Delhi Bombay Calcutta Madras Karachi
Petaling Jaya Singapore Hong Kong Tokyo
Nairobi Dar es Salaam Cape Town
Melbourne Auckland
and associated companies in
Berlin Ibadan

Oxford is a trade mark of Oxford University Press

Published in the United States
by Oxford University Press, New York

First published 1988
Reprinted (with corrections) 1990

British Library Cataloguing in Publication Data
Thomas, Pete
Abstract data types: their specification,
representation and use. − (Oxford applied
mathematics and computing science series).
1. Computer systems. Software. Design.
Application of abstract data types
I. Title II. Robinson, Hugh III. Emms, Judy
005.1'2
ISBN 0-19-859663-4
ISBN 0-19-859668-5

Library of Congress Cataloging in Publication Data
Thomas, Pete (Peter G.)
Abstract data types.
(Oxford applied mathematics and computing science series)
Bibliography: p.
Includes index.
1. Abstract data types (Computer science)
I. Robinson, Hugh. II. Emms, Judy. III. Title. IV. Series
QA76.9.A23T46 1988 005.7'3 88-5236
ISBN 0-19-859663-4
ISBN 0-19-859668-5 (pbk.)

Printed in Great Britain
by Bookcraft (Bath) Ltd
Midsomer Norton, Avon

Contents

About this book

This book is about *abstract data types*. In a few short years this topic
has moved from the frontiers of research to the foundations of
computer science. Ideally, we feel, abstract data types should form part
of the core teachings in computer science. Like many fast growing
areas, however, it takes a while for new material to become
incorporated into the mainstream syllabus of universities and
polytechnics. This book has been designed, therefore, for use in an
introductory course on data structures or programming, or to
supplement existing courses in these areas. It is appropriate to both first
and second years of study. The book is aimed primarily at students. To
do this, we have used techniques more commonly employed in the
construction of distance learning texts.

Abstract data types are used in almost all stages of software
development, particularly in specification, design, and implementation.
Chapter 1 begins by identifying the stages of software construction by
discussing the so-called *software life-cycle*.   There we show why it is
important to separate the specification of a piece of software from its
design and subsequent implementation. Specifying software in terms of
the data to be processed, and the operations that must be performed on
the data, has proved to be an excellent way of producing reliable
software and of reducing costs. Simply stated, an abstract data type is a
collection of data together with the operations that can be carried out on
that data.

These ideas are illustrated in *Chapter 1* by giving a non-technical
description of the abstract data type called stack. By the end of the first
chapter you should have a good understanding of what an abstract data
type is and how it fits into the specification stage of software
development.

A specification is a technical document primarily meant for use by
software designers. There is, therefore, a need to develop a language of
discourse by which specifiers of software can communicate to
designers. In essence, this language is highly structured, i.e. formal, and

is used for defining abstract data types. *Chapter 2* discusses two ways of formally defining the abstract data type stack. Our objective is to introduce you to two common approaches to specification in a familiar context.

In this book we have concentrated on developing specifications of well-known data structures: of stacks, queues, and trees because these will enable you to place the new material in readily assimilated contexts. We shall redress the balance, in *Chapter 8,* by looking at some less familiar applications.

Chapter 2 introduces the two specification methods. They are known as the *axiomatic approach* and the *constructive approach.* The axiomatic approach specifies an abstract data type by defining its properties as a set of *axioms.* An axiom is a generally accepted truth or principle and it is often the case that a small set of axioms is sufficient totally to define an abstract data type. Axioms are useful because they enable us to deduce further properties of the object. The constructive approach, as the name suggests, builds abstract data types from objects that are already well defined. In practice it is often the case that common abstract data types, such as those discussed in this book, are defined axiomatically and are then used in the constructive approach for more complicated applications.

Our first look at implementation and the role that abstract data types play occurs in *Chapter 3*. There you will see how the specification of an abstract data type, *stack* , can be converted to an implementation; we also provide some general guide lines for this conversion. In particular, we give alternative implementations to reinforce the idea that a correct specification should enable you to choose whatever design or implementation of the abstract data type that you like. The fact that abstract data types are used as the basis of a specification does, however, influence the way in which we approach the design. The implementations given in *Chapter 3* use UCSD Pascal. We have assumed that you already know how to program in Pascal, but if you have not met the additional facilities of UCSD Pascal — particularly *units* — you need not worry since they are fully explained in the text.

Chapter 4 investigates the facilities that a programming language should have in order to support abstract data types. In particular we look at encapsulation, a mechanism for separating out the specification of a subprogram (as indicated by its parameters) from the body of the subprogram. Such a mechanism is useful because it enables a programmer to use a subprogram without knowing the details of how it achieves its results (the algorithm). As the details of the algorithm (both the code and the local variables) are hidden from the user the technique is often referred to as *information hiding*. Encapsulation is the programming language analogue of an abstract data type! To illustrate this concept we look at packages in Ada and modules in MODULA-2. These languages can be viewed as (significant) extensions of Pascal and we use this fact when discussing their new features.

It will be useful here to say a little about our approach to implementation. We have attempted to separate the choice of data structure from the way in which subprograms are written to implement abstract data type operations. The choice of data structure is called the *representation,* whereas the subprograms that manipulate the data structures are referred to as the *implementation.*

Chapters 5, 6, and 7 have the same structure: they look at the specification, representation, and implementation of three of the most common abstract data types, *queue, deque* and *binary search tree.* Each of these abstract data types raises new problems in both specification and implementation.

Chapter 8 contains three case studies. These are examples of applications which are sufficiently realistic to convince you of the merits of our approach.

Finally *Chapter 9* shows you how the idea of an abstract data type can be used in program development along side such other established techniques as stepwise refinement. The history of this new approach is such that we tend to refer to *objects* rather than abstract data types (but they are essentially the same thing). The theme of this chapter is *reusable software* and the underlying idea is the same as the one you will meet in the constructive approach to specification. That is, rather than create a piece of software from scratch we use an existing piece of similar software and derive, or construct, the new software from it. In this sense the new software derives or *inherits* its properties from existing software.

We have included two kinds of exercise in this book: check points and exercises, and have included them at the most appropriate places for self study. The check points are simply a way for you to judge your own progress and consist of straightforward questions on the material just covered; their solutions should require little by way of written responses. Exercises, on the other hand, are substantial questions requiring more thought and will undoubtedly need a significant amount of time to complete. Our intention is that you should pause for a while at the check points to convince yourself that you are in good shape to continue. There should be little need to break off from studying the material whilst attempting the check points, so we have included their solutions immediately following in the text. With exercises we suggest that you answer them as they arise in the text, and break off from studying the text while you do so. A word of caution: as we sometimes develop teaching points in the exercises it is important for you to view their solutions as part of the main text. Solutions to the exercises are given at the end of the book.

At the end of the book we have included an annotated bibliography giving references to other important works in this area. You will find that there are a variety of different approaches to formal specifications, all of which offer insights into this important subject area.

Acknowledgements

This book is the result of ideas that began to develop during the making of the *Open University* computing course *M353, Programming and programming languages.* During the production of that course we came to realize that *abstract data types* are a fundamental and unifying concept in computer science and that there was a need for an introductory text in this area. We are indebted, therefore, to our colleagues at the *Open University* for many stimulating discussions, and also to those students who helped us formulate our ideas in a more usable manner. Thanks are also due to the faculty and students of the *University of Denver, Colorado, USA,* to whom one of the authors gave a one-quarter course based almost exclusively on the material in this book.

The Open University   P.G.T.
October 1987   H.M.R.
                                                                         J.M.E.

1 Abstract data types and software engineering

1.1 Introduction

In recent years the concept of an *abstract data type* has proved to be one of the most useful concepts in the design and construction of computer programs, and has helped to move software construction from the status of a craft to that of a profession. That this is so is evidenced by the fact that programming, in the widest sense, is now perceived as part of that engineering science known as *software engineering.* To understand the importance of abstract data types and the role of abstraction in the construction of software it is necessary to know what is meant by software engineering.

Commercial programs are often large and complex pieces of software that absorb huge amounts of effort in their production. It is most unusual for a single person to be involved in all stages of software production; indeed it is unlikely that any one person could fully comprehend the details of the construction of a complete modern software package. Since software is produced by teams of people, ways and means have to be found to organize and co-ordinate their tasks. In many disciplines it has been found that complexity can usefully be tackled through *modularity.* That is, complex tasks are subdivided into parts which are themselves often broken up until the pieces are sufficiently small to be handled by a single individual. Subdividing programs into procedures and functions is a technique with which we assume you are already familiar and, therefore, that you appreciate the benefits which accrue from this technique.

Within software engineering the term modularization usually means more than simply subdivision. It is a term that is applied to the process embodied within a program, and is a decomposition into *independent*

1

parts. That is, modularity is a procedural decomposition because it is
the method by which a program achieves its purpose that is
modularized. Latterly it has been observed that decomposing the data of
a problem into smaller subdivisions can have benefits equal to or
exceeding those of modularity of program code. Indeed, recent
advances in programming techniques seem to indicate that it is more
productive to examine the nature of the data in a problem since this
process can also reveal an appropriate modularization of the software.
This technique is known as object-oriented development and is a topic
that we shall pursue later in the book. For now we shall concentrate on
what is meant by subdividing the data and how abstract data types point
the way forward.

The design and construction process for software is itself divided
into subtasks, and there is general agreement that the stages shown in
Figure 1.1 provide appropriate subdivisions.

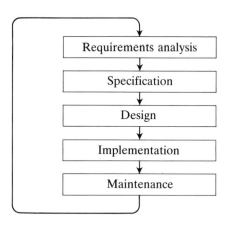

Fig. 1.1   The software life-cycle

The above figure illustrates the whole of the *software life-cycle* from
the initial idea for the particular piece of software, captured during the
requirements analysis stage, through to the time when the software
becomes obsolete and cannot be further maintained. Each of the stages
in the software life cycle are viewed as being distinct in the sense that it
is possible to define the start and finish of each stage. Output from one
stage, often in the form of a written document, forms the input to the
next stage. The process is iterative: the documents which form the

interface between stages can be incomplete, ambiguous, or even erroneous so that one stage cannot be completed without reference to one or more of the earlier stages. The whole process is cyclic because software ages and becomes obsolete, and the reasons for which the original software were written are no longer valid so the software must be redesigned. This means that the requirements must be re-evaluated and the cycle repeated.

The software life-cycle illustrates the close relationship of software engineering to other branches of engineering. Consider the task of building a motorway bridge across a river. The requirements analysis stage would identify basic characteristics of the bridge: its size — including length and number of traffic lanes; its strength — being capable of carrying a certain traffic load; the nature of the ground on which it is to be built; environmental considerations, and so on. All the requirements would then be turned into a technical document, the specification, that describes the characteristics of the proposed bridge using a language and form suitable for bridge builders. Such a document often forms the basis for tenders.

The specification of the bridge is also the input document for the design work. It is at this stage that models of the structure would be built and decisions as to the type of bridge would be taken. The design stage would produce blue-prints for the final construction (the implementation stage). For uncomplicated structures designers will often make use of tried and tested designs or will adopt standard components; the obvious reasons for this being savings in cost and the reliability of the final structure.

Finally there is the maintenance stage. Here, hopefully minor design faults are rectified and the bridge is kept in working order — we often say, 'kept up to specification', by which we mean that the bridge must not be allowed to decay to the point where it no longer matches its original specification. However, there is a more insidious problem. Traffic can increase beyond the level assumed in the requirements analysis thereby making the bridge unsuitable. It may be possible to strengthen the bridge by some skilful modification, but often the only real solution is to renew the bridge, and that means cycling back to the requirements stage again.

Dealing with complexity is an important topic in its own right, but there is often a more compelling reason to investigate the stages of (commercial) software development in greater detail — that of cost.

Software is an expensive commodity, particularly when compared to the cost of hardware which is becoming cheaper all the time. Software can take many expensive person-years of effort to produce because of its complexity, and that means it is difficult to produce a totally correct product. An aim must be, therefore, to devise methodologies and tools which can reduce the incidence of errors in the final product. That is, we want to be able to produce output documents from each stage that are *correct*.

The cost of software has two major components: (a) the cost of building the software in the first instance, and this includes the requirements analysis, specification, design, and implementation stages, and (b) the cost of maintenance; maintenance is what happens to the software during its lifetime as an actively used product.

There are two kinds of maintenance, one involving the correction of errors and the other involving changes due to revisions in requirements (software is often 'developed', during its lifetime to deal with new or revised situations that were not envisaged for the initial product). For what ever reason software is modified, the cost of maintenance has been found to far and away exceed the original cost of production. (*Fig. 1.2* compares the amount of a data processing department's budget spent on maintaining existing applications with the amount spent on devising new applications). The relative increase in cost of maintenance is the reason we search for software production techniques that can lead to fewer errors in the final product, and that also enable software to be easily (i.e. inexpensively) modified in later life.

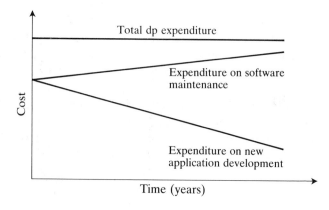

Fig. 1.2 Increasing cost of software maintenance

There is one area of software development which is now receiving much attention — software reusability. It seems an obvious remark to say that it makes economic sense to reuse a piece of existing software whenever possible since doing so will reduce the time and cost of software production. Also, if the existing software has been in use for some time, it will be more robust (contain fewer errors) than a newly created piece of software. It was not until recently that programming languages became available to make such reusability a realistic possibility.

Reusability is not, however, confined to the implementation stage alone. The fundamental idea is of considerable importance in the specification stage. We often produce specifications which apply to general situations and which can then be applied to specific problems. Such specifications are said to be *generic*. Generic specifications can lead to generic implementations — to programs that can be easily tailored to specific needs. Of course, for this to be a viable proposition programming languages must have facilities for enabling this process. We shall be discussing some of these features, and in so doing you will see both the adequacies and the deficiencies of current languages.

Specifications are important because they provide descriptions of software which are independent of implementation — the programming language and computer system on which the software will ultimately run. This means that the particular peculiarities or deficiencies of an implementation (areas in which errors often arise) can be ignored. In effect we produce specifications which are generic in the sense that they can be applied to any language and any computer. We usually choose to write specifications in a mathematical way in order to take advantage of well established fields of knowledge, enabling us to reason and argue about the specification in a way that gives us confidence in its correctness. In essence, formality means inventing a notation which is both succinct and avoids ambiguity. The notation must also have the property that it is easy to understand and manipulate (by software practitioners) so that, given some information about a piece of software — written in the notation — it is possible easily to derive additional properties of the software. This latter activity is usually called *reasoning* about software. Making specifications *formal* in this way also permits us to compare the final implementation with the specification and say whether or not the implementation is correct. This comparison is usually referred to as *validation*.

1.2   Specification

The specification stage is concerned with describing *what* has to be produced. Contrast this with the design stage which determines *how* a program will produce the required results. As a simple illustrative example suppose that you have the problem of rearranging a large set of names into alphabetical order. The *specification* of the program can be expressed in the following terms:

> *a program is required which will take as its input a set of names and will produce as its output the same names but in alphabetical order.*

Notice that the specification says nothing about how the re-ordering is to take place. There is no mention of the sorting method, nor of how the data is to be represented in the program. As far as the specification is concerned, as long as the output from the program is an ordered list of the input names the method of producing the output is *immaterial*. The advantage of removing the details of how the program produces its results (the implementation details) to the design stage is that the implementation can be altered at will (perhaps for reasons of efficiency an alternative sort method should be employed). Thus, the design can be changed *without altering the specification*. The issues of efficiency (speed of execution and storage space) are properly in the domain of the design process. As long as the design faithfully matches the specification then we can have confidence that the design is 'correct' in the sense that the design and the specification carry out the same task: they have the same *functionality*. There is also an intellectual advantage to separating out the 'what' from the 'how'. It reduces the complexity of the problem by removing unnecessary details and, as a result, we often describe this process as one of *abstraction*.

Notice also how the above specification concerned itself solely with the input and output of the program and the connection between them, i.e. that the output was an ordered list of the input data. This specification technique, in which we describe *what* is to be achieved by defining the input and output and their connection, will be used extensively throughout this book.

We shall now take our first look at an *abstract data type* and see how it can be described in terms of the above specification methodology. Our example will be the **stack** abstract data type. You are probably familiar with the use of stacks in computing; they turn up everywhere, an example is the one that stores the order of procedure invocations in recursive calls in Pascal. However, you probably think of a stack in terms of a data structure, for example, an array together with a variable that stores the index of the array element containing the item currently on top of the stack. Of course this isn't the only way of programming a stack, and you may be aware of other methods, using pointers, for example. Such views of stacks are, of course, *implementations* because they describe *how* stacks are represented in a program. What we need for a specification is a description of a stack that is devoid of implementation details and that does not depend on a particular programming language. In other words we want to capture the inherent properties of a stack so that, if we should ever meet one, we would immediately recognize it, no matter in what context it arose. What then characterizes a stack?

Here is one description of a stack. This description is not very precise but it does give the essential flavour of what it means to be a stack.

A stack is a collection of data kept in sequence. Each item of data is of the same type. Data is added to and removed from the sequence at one specific end of the sequence (usually called the top). It is possible to access only the top item of data in a stack.

In the above specification there is no mention of how the data is stored or how the operations of adding, deleting and accessing data are performed. *Fig. 1.3* shows a pictorial example of a stack:

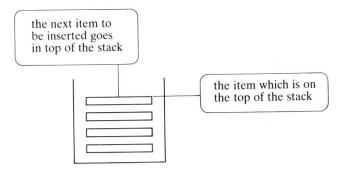

Fig. 1.3 A stack with four items inserted

The object in the figure *is* a stack yet it implies nothing about how one might go about programming such an object. This is exactly what we want from a specification.

The stack specification given above certainly gives the flavour of 'stackiness' but is rather imprecise for our purposes. It does, however, contain the essence of what a specification should contain. In particular it speaks of 'a collection of data' which is 'kept as a sequence' thereby describing a property that all stacks must have. The description also mentions three operations which involve stacks: adding a new data item, removing an item, and retrieving the data held in the top item of the sequence. You probably recognize these operations by their more familiar titles **push, pop,** and **top**. These operations involve only the top item in the sequence and serve to distinguish stacks from other abstract data types like **queues** (where operations can involve both ends of the sequence).

A fuller description of what a stack is consists of descriptions of these operations:

top *is an operation that returns the item at the top of the stack as its result;*

push *is an operation that, given an item and a stack, returns a stack with that item inserted at the top;*

pop *is an operation that takes a stack and returns, as its result, that stack with its top item deleted.*

It is usual to include two further operations for all abstract data types: one which tells whether or not there are any items in the abstract data type and another which creates a new, empty instance of the abstract data type. In the case of a stack we have:

createstack *is an operation which returns a new empty stack;*

isemptystack *is an operation which returns true, when the stack is empty, or false when the stack is not empty.*

There are three aspects of these descriptions which are fundamental to the descriptions of all operations regardless of the abstract data type:

(i)   there are values passed to the operation — the *source data* — for example, an item and a stack in the case of **push**;

(ii) there are values passed from the operation, as the *result* of the operation having been carried out;

(iii) the transformation of source data into results: that is, the relationship that the results have to the source data. (Note that this does not imply anything about the method by which the transformation is achieved.)

A way of visualizing an abstract data type is as a 'black-box' which offers you choices of operations that define the abstract data type; all you have to do is select the operation and provide the correct source data, and the black-box will do the rest. How the results are obtained by the black-box is immaterial. *Fig.1.4* shows a pictorial representation of the stack abstract data type.

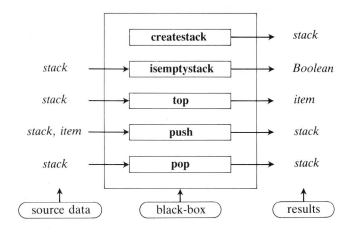

Fig. 1.4   The stack abstract data type viewed as a black-box

We can now say that a stack is defined by the five operations outlined above; any object that behaves in a manner consistent with these operations counts as a stack.

You may well have seen stacks differently defined. For example, it is common to omit the **top** operation and define **pop** as an operation which returns both the item at the top of the source stack *and* the source stack with the top item deleted. Such a definition is clearly different from the one we have been using and, therefore, describes a *different* abstract data type, albeit with the same name. Hence, a set of operations defines an abstract data type, and a different set of operations (no matter how small the difference) constitutes a different abstract data type.

It should be clear to you that it is a quite straightforward matter to define the source data and the results of operations without knowing how the former are transformed into the latter. Precisely how the transformation is described forms the bulk of this book. For now, we conclude this section with a taste of how such descriptions can be achieved. Observe that the meanings of the five stack operations are not independent. To perform any operation on a stack that stack must first have been brought into existence with **createstack**. Once in existence the operation **push** can be applied to the stack in order to insert items into it. Only after items have been inserted into the stack is it sensible to apply the operations **top** and **pop**. Of course, the result of applying **isemptystack** depends on the history of operations applied to the stack.

It is possible to relate the meanings of the stack operations explicitly. For example:

if   an item is **push**ed onto an existing stack
                         and
         the operation **top** is applied to that stack
then
         the item that is returned is the item previously **push**ed on
         to the stack.

Similarly,

if   a new stack is created (using **createstack**)
                         and
         the operation **isemptystack** is applied to that stack
then
         the result will be the value *true*.

These two relationships have an important property in common — they are *always* true no matter what state the stack may be in. They are therefore examples of *axioms*.

You may be tempted to think that the descriptions given above are quite sufficient for our purposes. We have been able to describe an abstract data type by defining its operations in a way that is independent of any representation or implementation. This is almost true. Unfortunately our informal style lacks precision and is not very useful for manipulative purposes. There are many questions about stacks that our descriptions fail to answer. For example, is it legal to apply the operation **top** to an empty stack and, if so, what should be the result? It is not simply a question of finding answers to such questions — after all we are defining what *we* want a stack to be and we can choose whatever answer suits us best. What *is* required, however, is that we *fully and unambiguously* define all the operations in such a way that anyone else wanting to discover what we mean by a stack can do so by reference to the specification alone.

Check point

(i)   What is an abstract data type?

(ii) What is the major difference between a specification and an implementation?

(iii) What are the three components of the definition of any operation?

(iv) What are the operations that define the abstract data type stack as we have described it?

Solutions

(i)   An abstract data type is a collection of related operations. The behaviour of the abstract data type can only be seen by observing the results of applying the operations.

(ii) A specification is a definition of an abstract data type that is devoid of implementation details. An implementation is the production of a computer program which carries out the operations defining an abstract data type.

(iii) Source data, results, and the relationship between them.

(iv) There are five operations: **createstack, isemptystack, top, pop,** and **push**.

2 Formal specification of abstract data types

2.1 Introduction

In *Chapter 1* we introduced the idea of defining an abstract data type as a *collection of related operations*. In this chapter we will examine the problem of *precisely defining* these operations to give a *formal specification*. In subsequent chapters we will concentrate on features of programming languages, particularly of the language Pascal, which enable us to develop *representations* and *implementations* of these specifications.

Chapter 1 concluded by remarking that the operations that characterized a stack had been defined only in an *informal* fashion, that is, the descriptions lack *precision* in their meaning. Precision is needed in a specification, as it is a prerequisite for the achievement of correctness (and all the consequent benefits that accrue later in the software life cycle). Precision can be achieved by expressing a specification in a formal language based on mathematics. A formal language is a language with a previously agreed set of rules governing its use, such that a statement in the language has *exactly one meaning* for the community making use of the language.

Formal languages are sometimes contrasted with the so-called *natural languages* of normal human discourse, such as English. The contrast is usually made on the grounds that natural languages are imprecise by virtue of the fact that they supposedly allow the construction of statements that have more than one meaning. A typical example offered to support this assertion of ambiguity is the joke *'Time flies. You can't; they fly too fast.'* The joke depends on *'Time flies'* being interpreted as two grammatical structures. In the intended sense, *time* is a noun and *flies* is a verb, whereas the joke takes *time* as a verb and *flies* as a noun. However, such ambiguity is more apparent than real. The fact that

12

we can recognize the joke for what it is demonstrates the preciseness of natural language: we regard the statements as a (not very funny) joke and not as an example of serious ambiguity. Natural language *can* be used with precision, and with a precision that has an infinite richness. However, this precision is achieved within a particular social group by the continuous *negotiation* and *construction* of meaning: there is a set of previously agreed rules which is continuously changed and augmented as normal social discourse proceeds (see Wittgenstein, 1958; Walker and Adelman, 1976, for example). Paradoxically, it is this very richness of precision in natural language, this process of negotiation and construction of meaning, that makes it unsuitable as a vehicle for precise specifications in software engineering. Precise natural language specifications for the software engineering community could be produced, but it is likely that they would require an inordinate number of words to lay bare the assumptions and rules about meaning that are being employed. The length and verbosity of such specifications militates against clarity and understanding. Particularly, natural language specifications are difficult to use when *reasoning* about an abstract data type in order to, say, predict additional characteristics of its behaviour. In true contrast, a *formal language* offers a *lingua franca* for the software engineering community, based on mathematics, where the assumptions and rules about meaning are already given and fixed. There is no need to negotiate and construct meaning within the limited context in which such a formal language is used for specification purposes. As a consequence, a specification expressed in a formal language will be both precise and (relative to natural language) brief. It will also be a language that can be used to *reason* about the behaviour of the object being specified. Of course, the precision and succinctness of a formal language makes it unsuitable for the expression of jokes, unfunny or otherwise.

In all likelihood, you are already familiar with at least one formal language, in the sense that you know a programming language (hopefully, given the intent of this book, the language Pascal). Any programming language is a *formal language,* since a program written in that language has a single, fixed meaning. Ultimately, this meaning is defined by the combination of compiler and machine used to execute the program. Repeated executions of the program yield the same result, given the same input, if any, to the execution. However, although programming languages are formal languages, they are not the kind of formal language that will be of use in the specification of an abstract data type. Essentially, this is because they are languages for describing a *representation,* not an *abstraction.* Consider, for example, an operation, **maximum**, which we wish to define. We can informally describe **maximum** as follows.

> **maximum** *is an operation that takes two integers as source data and returns the integer with the higher numerical value as the result (or either integer if they are equal).*

If we specify this operation in the formal language Pascal, we might obtain the following:

```
function Maximum (x, y: integer): integer;
begin
  if x >= y
  then
    Maximum := x
  else
    Maximum := y
end;
```

As a formal specification, this Pascal function fails to suppress the representation details. Although it does define precisely the transformation of the source data into the result, it does so by means of an algorithm that details *how* the transformation is achieved, not *what* is required of the transformation. We could, for example, replace the body of the function by other equivalent statements without changing its implicit meaning.

Exercise 2.1
Change the body of the function *Maximum* so that it is different from the original version yet achieves the same result.

To summarize: we cannot define a *representation-free* specification by means of a language the *purpose* of which is to describe representations. We require a different formal language: one that encourages abstraction in order to express ideas about *specification* distinct from *representation.*

2.2 Syntax and semantics

This chapter will describe two approaches to the formal specification of abstract data types: the *axiomatic* approach and the *constructive* approach. The reasons for adopting *two* approaches will be made clear later. However, before beginning a detailed description of these approaches, an important distinction needs to be established: the distinction between *syntax* and *semantics*. This distinction can be illustrated by means of the description of the *stack operations* introduced in *Chapter 1*. The five operations described were: **createstack, top, pop, push,** and **isemptystack**. If we look at the description of each operation, we can see that it involves a description both of *syntax* and of *semantics*. For example, the **isemptystack** operation was described as

> **isemptystack** *is an operation which returns true, when the stack is empty, or false, when the stack is not empty.*

A language (such as English or Pascal) consists, in its elementary parts, of a vocabulary and a set of punctuation symbols (the words and punctuation of the English language, the keywords, allowable identifiers and punctuation in Pascal). Elements of the vocabulary and the punctuation set are juxtaposed to form *expressions* (such as sentences in English and programs in Pascal). The syntax of a language specifies the legal juxtapositions. An expression which adheres to the *syntax* of the language in which it is expressed is said to be *syntactically correct*. Thus, **isemptystack** has a *syntax* of a stack as source data and a Boolean value as the result. That is, a *syntactically correct* use of **isemptystack** requires that it takes a stack as source data and returns a Boolean value as the result. In contrast, the *meaning* of the various forms of expression in a language is called the *semantics* of the language. In the case of specifying an operation, the semantics define the relationship between the result and the source data. The *semantics* of **isemptystack** are that the result is the value *true,* when the source data stack is empty, and the value *false,* when the source data stack is not empty.

Exercise 2.2

Give the syntax of the **pop** and **push** operations, as described in *Chapter 1*. That is, for each operation, describe the types of the source data and the result.

Exercise 2.3

Give the semantics of the **pop** and **push** operations as described in *Chapter 1*.

We can apply this distinction to the problem of formally specifying the operations that define an abstract data type, since it allows us to refine the problem into two sub-problems:

(i) How do we specify the syntax of operations (that is, the nature of the source data and result) in a formal language?

(ii) How do we specify the semantics of operations (that is, the meaning) in a formal language?

This refinement of the problem is mirrored in its solution: the formal specification of an abstract data type involves a specification of the *syntax* of the operations and a specification of the *semantics* of the operations. However, these are not the only two entries in the formal specification. A necessary precursor to any syntax and semantics is a *vocabulary*, declaring the linguistic objects that appear in the definition of the syntax and semantics. The vocabulary, syntax, and semantics describe an abstract data type, which itself must be formally declared by means of a *name*. Thus, there are four entries in a formal specification of an abstract data type:

A *NAME;*
The *SETS* of objects manipulated by the operations;
The *SYNTAX* of the operations;
The *SEMANTICS* of the operations.

Although this chapter describes two approaches to formal specification, the *NAME, SETS,* and *SYNTAX* entries have the same form for both approaches. It is only in the *SEMANTICS* entry that the approaches differ.

We can illustrate the first two entries by means of the example of a stack. As the *NAME* entry we have

NAME
    stack (item)

which states that the abstract data type *stack* is a collection of *items*. As can be appreciated, there are many uses of a stack. Applications exist that require stacks of *numbers*, stacks of *employee names,* stacks of *run-time data* during the execution of a program, etc. These various stack applications are clearly all different, in the sense that what is being placed on the stack in each case is different. Yet, they are all similar, in the sense that they are all *stacks* (provided they all make use of just the same five characterizing stack operations). That is, they can all be **pop**ped, **top**ped, **push**ed, and so on. It would therefore be useful if a formal specification of the abstract data type stack could be developed that captured the similarity (the 'stackiness'), yet ignored, for the moment, the differences (the nature of the items placed on the stack). This is precisely what our *NAME* entry begins to do: it is the first entry in the *generic* specification of a stack. For example, suppose we wish to specify a stack of *employee names,* where the order of placement on the stack is the order in which the named employee joined the company. The generic *NAME* entry becomes

NAME
    stack (employee name)

and similar replacements are made throughout the rest of the specification (as you will see). The declaration of **stack** (item) therefore allows us to develop a *generic* specification for a stack that provides a *template* for the specification of particular *applications* of that stack.

The *SETS* entry declares the sets from which the particular values manipulated by the operations of the abstract data type are drawn. So, for the stack, we have the following declarations:

SETS
  S   set of stacks
  I   set of items
  B   set consisting of the Boolean values *true* and *false*

That is, the vocabulary that will be used in the definition of the syntax and semantics of the operations consists of members of the set *S,* members of the set *I,* and members of the set *B*. Within this vocabulary, the set *S* has a definitional status distinct from *I* and *B*. The specification of the abstract data type stack defines ultimately what it is to be a member of *S;* it defines (amongst other things) how members of *S* (stacks) may be legitimately constructed. In contrast, the specification does not address such issues for the sets *I* and *B:* it does not describe how items and Boolean values may be constructed. The sets *I* and *B* are assumed to have been defined elsewhere. This distinction is sometimes effected by referring to *S* as the *carrier* set and to *I* and *B* as *auxiliary* sets.

2.3 Defining syntax

We use *SETS* in the formal definition of the syntax of operations. So far, our informal description of the syntax of **pop** has been:

> *the syntax of* **pop** *is that its source data is a stack and its result is a stack.*

We shall express the syntax of an operation as

> ***operation name:*** *source data* → *result*

Thus the syntax of **pop** is represented formally as:

> **pop:** S → S

There are three components of this formal definition of syntax.

(i) The declaration of the *name* of the operation — in this case **pop**.

(ii) The declaration of the *type of source data* for the operation. This is done by giving the name(s) of the set(s) involved. The symbol **:** is used as a delimiter, to separate the declaration of the name of the operation from the declaration of the source data. In this case, *S* declares that the source data is any one stack from the set of stacks.

(iii) The declaration of the *type of the result*. This is done by giving the name(s) of the set(s) involved. The symbol → is used as a delimiter, to separate the declaration of the source data from the declaration of the result. In this case, *S* declares that the result is one stack from the set of stacks.

So, *pop: S* → *S* can be read as: *pop takes any member from the set S (that is, any stack) as source data and produces a member from the set S (that is, a stack) as the result.*

The syntax of the **top** operation is

top: S → I

That is, **top** is an operation that takes a stack as its source data and returns an item as its result.

We should emphasize what is meant, for example, by the set *S* appearing in the source data component of a syntax definition. Its meaning is that an application of the operation can take any one member from the set *S* as its source data. It does not, of course, mean that an application of the operation takes the set S (which may have an infinite number of members) as its source data. Similar remarks apply to the entries in the result component.

These syntax definitions for **top** and **pop** illustrate an important point concerning the formal specification of the syntax of an operation. Any value that is *changed* or *created* as a result of an operation, appears, by virtue of the name of the set to which it belongs, in the result component of a syntax definition. Hence, the syntax definitions make it quite clear that **top** does not change the stack taken as source data: the set S does not appear in the declaration of the type of result. Since S is not listed as part of the result, no change to the source data stack takes place. Therefore, we have introduced some of the precision required with the formal specification. It may be recalled from *Chapter 1,* that the informal description of **top** was imprecise as to whether or not a stack was produced in the result. The unchanged source data stack still *'exists';* it is merely not made available as part of the result of **top**. Similarly, the syntax definition of **pop** makes it clear that **pop** does *not* return an item as part of the result, even though it deletes an item. However, the syntax definition says nothing further concerning the *meaning* of **top** or **pop**. For example, what *cannot be inferred* from these syntax definitions is that if a stack appears in the source data *and* in the result, then it must be changed by the operation. We can only infer that if a stack appears in the source data and does *not appear* in the result, then it is *not changed* by the operation. This may seem a very arbitrary and complex rule to apply to syntax definitions, but it is, in fact, very useful in achieving precision. Whatever an operation does, even if it merely copies its source data as its result, must be displayed in its syntax definition. The syntax is clear and precise and allows no possibility of ignoring any effects of the operation. As an illustration, suppose there was a requirement for an operation, named (say) **extratop**, which was equivalent to **top**, except that the unchanged source data stack was also returned as part of the result; then the syntax definition would be

extratop: $S \rightarrow I \times S$

Now the result is defined as a member of the set obtained from the Cartesian product (denoted by the operator $\times$) of I and S; that is, the result is a *pair,* consisting of an item *and* a stack, in that order. We cannot infer from this definition that **extratop** changes the stack in any way, just because it appears in the result component.

Exercise 2.4

Suppose that a new operation involving stacks is to be specified. The operation is named **replicatestack** and takes a stack as source data and produces an identical stack as the result. Write down a formal definition of the syntax of **replicatestack**.

Exercise 2.5

Write down a formal definition of the syntax of the **push** operation described in *Chapter 1*.

Notice, in the solution to *Exercise 2.5,* that the order of the source data sets is significant: the source data for an application of **push** is a pair, consisting of an item and a stack, in that order. A pair consisting of a stack and an item, in that order, would not be valid source data. You will see, in the solution to *Exercise 2.7,* another example of this significance of ordering.

Exercise 2.6

Suppose that a new operation involving stacks is to be specified. The operation is named **mergestacks**, and takes two stacks and merges them (the semantics of this merging are irrelevant to the exercise) into one stack. Write down the formal definition of the syntax of **mergestacks**.

Exercise 2.7

Write down the syntax of an operation, **subtract**, that takes as its source data two integers and produces as the result the integer obtained by subtracting the second source integer from the first source data integer. Assume that the set of real numbers is denoted by Z.

Exercise 2.8

Write down a formal definition of the syntax of the **isemptystack** operation described in *Chapter 1*.

The syntax of the final operation, **createstack**, is

   createstack: $\rightarrow$ S

Since **createstack** takes no source data, the corresponding component is omitted from the syntax definition.

So far, we have assumed that all five of the operations will produce results which are members of either *I, S,* or *B.* A closer examination of one operation, **top**, shows that this assumption is not well-founded. The syntax of **top**,

top: $S \rightarrow I$

states that its result is an item. However, **top** when applied to an empty stack cannot sensibly produce an item as the result. Such an application of **top** is sometimes called an *error situation.* For example, **top** may be so defined that its application to an empty stack is regarded as producing an *'error'. 'Error'* is something of a misnomer: it would be more correct to say that, in the case of applying **top** to an empty stack, the result is not an item (a member of *I*) but some *indication* that it is not sensible to return an item as the result in these circumstances. It is not an 'error' or 'mistake' to apply top to an empty stack. It is merely a situation in which it is appropriate to return a value *other* than an item as the result. In this chapter, we shall assume that it will be appropriate to return a *message* as the value of the result. It is possible to emphasize the 'unexceptionality' of returning a *message* by incorporating the *message* value — say, the string *the stack is empty* — as a member of the set *I* (see Martin, 1986, for example). **top**ping an empty stack then always returns a result from *I* but, in one case, the value indicates that an empty stack has been **top**ped. We prefer, however, to indicate that a *message* is different from an *item* and declare this appropriately. Hence, the *SETS* entry would contain the declaration:

M   set of message values

M has one member denoted by the string *the stack is empty.* The syntax of **top** then becomes:

top: $S \rightarrow I \cup M$

Here the result is defined as *one* object from the set obtained from the union (denoted by $\cup$) of *I* and *M*. That is, the result can be *either* a member of *I or* a member of *M* (but not both, since $I \cap M = \varnothing$).

The only other operation where a similar problem *may* arise is **pop**. What is the result of **pop**ping an empty stack? We shall assume, for the moment, that **pop**ping an empty stack results in an empty stack, so there is no need to change the syntax of **pop**.

Exercise 2.9

What is the difference between the type of result specified by the following two syntax definitions?

top: $S \rightarrow I \cup M$

anothertop: $S \rightarrow I \times M$

It is useful to make a distinction between such situations that arise *inherently* and those that arise as the result of a *representation*. For example, a stack can be represented by means of an *array* in Pascal. An array in Pascal is a *finite* data structure, that is, the number of elements is fixed. A stack is a potentially infinite data structure, that is, the number of items is not fixed and can vary up to an arbitrary large number. As a consequence, an array representation of a stack will necessarily involve a constraint on **push** when the stack is full (that is, all elements in the array have been used for items) purely because of the representation, and it may be deemed appropriate to return a message value when this constraint is exceeded, since it is clearly not possible to return the expected stack. Such situations can be contrasted with the account given above of **top**ping an empty stack, which is seen as somehow inherent to the specification of a stack. The distinction can, however, be more apparent than real on occasion. It can be necessary to include in the specification some constraint on the number of items in a stack *because* of the logic of the stack application. For example, in order to specify a stack of employees, where the rules of the company state that no more than 100 people may be employed at any one time, a constraint must be captured at the specification stage and must be recorded in the formal definition of the syntax and semantics of the appropriate operation.

Exercise 2.10

The following declarations describe an application requiring a stack of employees. How should they be modified to incorporate the constraint that no more than 100 people may be employed?

NAME
   employeestack(employee)

SETS
   P   set of employeestacks
   Y   set of employees
   B   set consisting of Boolean values *true* and *false*
   M   set of message values, consisting of the single member *the employee stack is empty*

SYNTAX
   createemployeestack: $\rightarrow$ P
   topemployeestack: P $\rightarrow$ Y $\cup$ M
   popemployeestack: P $\rightarrow$ P
   pushemployeestack: Y $\times$ P $\rightarrow$ P
   isemptyemployeestack: P $\rightarrow$ B

With these asides on the incorporation of *message* values, it is useful to give the syntax definitions of the five stack operations together, as in *Fig. 2.1.*

SYNTAX
   createstack: $\rightarrow$ S
   top: S $\rightarrow$ I $\cup$ M
   pop: S $\rightarrow$ S
   push: I $\times$ S $\rightarrow$ S
   isemptystack: S $\rightarrow$ B

Fig. 2.1 The syntax of the five stack operations

Check point

(i)   What does the following declare about the syntax of **anyop**, where *S* and *I* denote the sets previously defined?

>   **anyop:** $S \rightarrow I \times S$

(ii)   What does the following declare about the syntax of **anotherop**, where *S, I,* and *M* denote the sets previously defined?

>   **anotherop:** $S \times I \rightarrow (S \times B) \cup M$

Solutions

(i)   The syntax of **anyop** is that its source data is a stack and its result is a pair consisting of an item and a stack, in that order.

(ii)   The syntax of **anotherop** is that its source data is a pair consisting of a stack and an item, in that order, and its result is either a pair consisting of a stack and a Boolean value, in that order, or the message *the stack is empty.*

We now turn our attention to using these syntax definitions in the formal specification of the semantics of operations.

2.4   Defining semantics axiomatically

2.4.1   *Introduction to the approach*

As we have mentioned, we are describing two approaches to the formal specification of abstract data types: the *axiomatic* approach (also known as the *algebraic* approach) and the *constructive* approach (also known as the *operational* approach). In a sense, the axiomatic approach is the more fundamental, since the operations of an abstract data type are defined by relating their meanings to one another, *without* reference to any operations other than those characterizing the abstract data type. In contrast, the constructive approach defines the abstract data type operations *in terms of* some other known set of operations that are not those characterizing the abstract data type. These *other operations* form the *underlying model,* upon which the definition of the abstract data type is built, or *constructed.* Obviously, if the abstract data type operations are to be defined in terms of the operations of an underlying model, then this underlying model needs to be defined in some fashion. The definition of the underlying model can be *implicit,* as is the case where the underlying model operations are those of accepted mathematical knowledge, such as set theory. In such a case, it is assumed that a formal definition of operations on sets could be given, if necessary, but since all practitioners agree on the precise meaning of operations on sets, this is not necessary. Alternatively, the underlying model is defined *explicitly,* using some formal language. We shall take this latter alternative of explicitly defining the underlying model, by means of the formal language available to us: that of the axiomatic approach. Such a dependence of the constructive approach on the axiomatic approach is, as we shall see, extremely useful. In essence, a general underlying model can be defined axiomatically and then used as the basis for the constructive specification of a range of related abstract data types. Further details of specification techniques are given by Liskov and Zilles (1975), Guttag, Horowitz, and Musser (1978), and Jones (1980, 1986).

2.4.2   The basic approach

We now turn our attention to the axiomatic approach where the operations are implicitly defined by *axioms* which relate the meanings of operations to one another. Before becoming involved in the detail of the axiomatic approach, it is useful to clarify what we mean by *relate the meanings of operations to one another.* We have, in fact, already illustrated this in *Chapter 1,* when describing the 'stackiness' of stacks. There it was noted that the meaning of the operations that characterize an abstract data type were not independent of one another. The example used is shown here as *Fig 2.2.*

if   an item is **push**ed on to an existing stack
                              and
        the operation **top** is applied to that stack
then
        the item that is returned is the item previously **push**ed on
        to the stack

Fig. 2.2

In *Fig. 2.2,* the meanings of **push** and **top** are related to each other. A further example is given in *Fig. 2.3,* where the meanings of **createstack** and **isemptystack** are related to each other.

if   a new stack is created by **createstack**
                              and
        the operation **isemptystack** is applied to that stack
then
        the value *true* is returned

Fig. 2.3

These two examples relate the meanings of operations to one another by asserting something about the application of one operation to the result of applying another. They are both examples of *axioms,* albeit expressed in (stilted) English rather than a formal language. As was noted in *Chapter 1,* an important characteristic of an axiom is that it is *always* true. Thus, the way in which *Fig. 2.2* relates the meanings of **push** and **top** is *always* true, no matter what the state of the stack may be or the value of the particular item used.

In order to express such axioms in a formal language we need a notation for expressing the application of an operation to particular source data values. Using **push** as an example, its application to a particular stack and a particular item, that is, to particular instances of the source data sets, is shown by

push(i, s)

Here the source data are denoted by *i* and *s,* standing for a specific instance of an item and of a stack, respectively. By convention, the order of the instances is the same as the order of the sets in the syntax definition. This is usefully reinforced by using the lower case version of the upper case letter used in the syntax definition. Where the application of an operation does not require any source data (from the syntax definition), an application of the operation is shown in a similar fashion, omitting any reference to source data values. For example

createstack

denotes an application of the **createstack** operation.

This notation can now be used to express formally the way in which the meanings of **push** and **top** are related. The example of *Fig. 2.2* can be expressed formally as the following *axiom* (which is numbered for ease of reference):

top(**push**(i, s)) = i   *(S1)*

The axiom involves applications of the two operations **top** and **push**. First, **push** is applied to the source data values, *i* and *s*. That is, the item *i* is **push**ed onto the stack *s*. This is denoted by

push(i,s)

nested within the outermost pair of brackets. The result of this application of **push** will be a stack; this is known from the syntax definition of **push**. **top** is then applied to this result stack. That is, **top** is applied to the result of **push**(i, s) denoted by

top(**push**(i, s))

This application of **top** yields an item as its result; this is known from the syntax definition of **top**. *Axiom S1* completes matters by stating that this result item will be the same as the item originally **push**ed onto the stack, since *i* denotes a specific instance of a stack.

In a similar fashion, the example of *Fig. 2.3* can be expressed formally as

isemptystack(**createstack**) = *true*   *(S2)*

Axiom S2 states that whenever an application of **createstack** (which yields an empty stack as its result) is used as the source data for an application of **isemptystack,** then the result (which is of type Boolean, from the syntax definition) is always the value *true.*

The use of *whenever* in the description of *axiom S2* emphasizes the important general point about axioms: they are true for any particular values that adhere to the syntax definition of the operations involved. That is, *axioms S1* and *S2* are true for *all* stacks and for *all* items. We should, therefore, preface the two axioms with such a statement as 'No matter what the value of *i,* provided it is a member of the set *I* (of items), and no matter what the value of *s,* provided it is a member of the set *S* (of stacks), the following axioms are always true.'

Further axioms are now developed which relate meanings of the remaining operations of a stack to one another. The resulting axioms will together define the semantics of the abstract data type stack. We have already seen that

push (i, s)

denotes the application of **push** to an item and to a stack. This application of **push** yields a stack as its result, that is, it evaluates to a stack. The precise nature of this result stack depends on what the source stack, *s,* contained. It is certain, however, that even if *s* were empty, the result would be a stack containing at least one item. In other words,

push (i, s)

always evaluates to a stack containing at least one item. Hence, if the **isemptystack** operation is applied to this stack, the result will always be *false.* This can be expressed as an axiom that relates the meaning of **push** to **isemptystack,** thus

$$\textbf{isemptystack}(\textbf{push}\,(i,\,s)) = \mathit{false} \qquad\qquad (S3)$$

Axiom S3 can be read as 'no matter what value *i* has and no matter what the state of *s* is, whenever the item is **push**ed on to the stack the resulting stack is never empty'.

The next axiom expresses the way the meanings of **pop** and **push** are related. It is

pop(**push**(i, s)) = s   *(S4)*

That is, no matter what value *i* has and no matter what the state of *s* is, if the item is **push**ed onto the stack and the stack is then **pop**ped, the result is the original stack.

It is worth noting how the axioms are making the description of the semantics precise. The syntactic definition of **pop** was

pop: S → S

which, apart from the operation name, has the same syntactic structure as the solution to *Exercise 2.4,* that is

replicatestack: S → S

In *Section 2.3,* the point was made that if a stack appears in the source data declaration and also appears in the result declaration, then it cannot be inferred that the result stack is necessarily different from the source data stack. Whatever an operation does, even if it merely copies its source data as its result (as is the case with *replicatestack*), is displayed in the *syntax* definition. It can now be seen from the *semantics* in one set of circumstances, that **pop** both produces a result stack different from the source data stack and that it states specifically what the difference is. Intuitively, the situation in which **pop** does produce a result stack different from the source data stack arises when the source data stack is not empty. We have already seen, in the development of *axiom S3,* how we can produce a non-empty stack as the source data for an operation; it is achieved by

IT GOES DOWN HILL AFTER THIS !!!

push(i, s)

Thus, *axiom S4* applies **pop** to this non-empty stack and specifies the result that will be obtained. We now have to deal with the situation where **pop** is applied to a stack that is empty. This application can be written as

pop(**createstack**)

The informal description of **pop** given in *Chapter 1* did not address this issue of what **pop** produces when applied to an empty stack. One benefit of using a formal language is that it can be used to *reason* about the behaviour of a specified object. We have begun to do that: the development of the syntax of **top** and **pop** (at the end of *Section 2.3*) focused attention on an overlooked issue: their action on an empty stack. That focus of attention is resumed in the development of the axiomatic specification. Of course, the formal language does not indicate *what* the action should be: it merely indicates the *need* for a decision on the part of the person producing the specification. In the autocratic role of authors, we have decided that **pop**ping an empty stack should produce an empty stack as the result.

Exercise 2.11
Write down the axiom that specifies the action of **pop** when applied to an empty stack.

Exercise 2.12
Write down an axiom that specifies the action of **replicatestack** when applied to a stack. Assume that replicatestack always produces a result stack that is a copy of the source data stack, irrespective of whether or not the source data stack is empty.

Notice how *axiom S4* (above) and *axiom S5* (given in the solution to *Exercise 2.11*) have precisely defined the semantics of **pop** and how, together with *axiom S6* (given in the solution to *Exercise 2.12*), they clearly define the semantic difference between two operations that are syntactically identical in their source data and result.

The utility of a formal language for the formulation of a precise specification can be further demonstrated by returning to *axiom S1*. There the meanings of **top** and **push** were related. It can now be appreciated that *axiom S1* defines the action of **top** in the circumstance of its application to a non-empty stack: this is guaranteed by the use of **push**(i, s) as the source data for the application of **top**. This suggests that we also need to consider the case of applying **top** to an empty stack. Recalling the discussion on messages at the end of *Section 2.3,* we choose that **top**ping an empty stack should yield a *message* value as the result.

Exercise 2.13

Write down the axiom that specifies the action of **top** when applied to an empty stack.

The solution to *Exercise 2.13* completes the axiomatic definition of the semantics. To summarize, we now give the full specification using the axiomatic approach below in *Fig. 2.4*.

NAME
  stack (item)

SETS
  S   set of stacks
  I   set of items
  B   set consisting of the Boolean values *true* and *false*
  M   set of message values consisting of the single member
       the stack is empty

SYNTAX
  createstack: → S
  top: S → I ∪ M
  pop: S → S
  push: I × S → S
  isemptystack: S → B

SEMANTICS
  ∀i ∈ I, ∀s ∈ S:
  top(**push**(i, s)) = i   *(S1)*
  isemptystack(**createstack**) = *true*   *(S2)*
  isemptystack(**push** (i, s)) = *false*   *(S3)*
  pop(**push**(i, s)) = s   *(S4)*
  pop(**createstack**) = **createstack**   *(S5)*
  top(**createstack**) = *the stack is empty*   *(S6)*

Fig. 2.4 The specification of the abstract data type stack

The assertion '∀i ∈ I, ∀s ∈ S' at the beginning of the *SEMANTICS* entry can be read as 'for all *i* that are members of the set *I,* and for all *s* that are members of the set *S* the following is true' (∀ being known as the universal quantifier). That is, this assertion is a more compact rendition of the generality noted after the discussion of *axiom S2:* 'No matter what the value of *i,* provided it is a member of the set *I* (of items), and no matter what the value of *s,* provided it is a member of the set *S* (of stacks), the following axioms are true'.

Exercise 2.14

Write down *axiom S5',* a modified version of *axiom S5,* required to define that **pop**ping an empty stack results in the message value *the stack is empty,* rather than an empty stack.

Since the next section makes use of this modified version of **pop**, *axiom S5'* is included with *axioms S1 – S4* and *S6* to give the *SYNTAX* (the syntax of **pop** has now, of course, changed) and *SEMANTICS* of a modified specification in *Fig. 2.5,* below.

SYNTAX
    createstack: → S
    top: S → I ∪ M
    pop: S → S ∪ M
    push: I × S → S
    isemptystack: S → B

SEMANTICS
∀i ∈ I, ∀s ∈ S:
top(**push**(i, s)) = i	(*S1*)
isemptystack(**createstack**) = *true*	(*S2*)
isemptystack(**push** (i, s)) = *false*	(*S3*)
pop(**push**(i, s)) = s	(*S4*)
pop(**createstack**) = *the stack is empty*	(*S5'*)
top(**createstack**) = *the stack is empty*	(*S6*)

Fig. 2.5 A modified specification

2.4.3   Completeness

Before concluding this section on the axiomatic approach, the issue of *completeness* needs to be addressed. Implicitly, a claim has been made that a precise definition of semantics has been produced. For example, it is an implicit claim that the axioms of *Fig. 2.5* precisely define a stack where, in particular, **top**ping and **pop**ping an empty stack results in a *message* value. One aspect of the claim that these axioms are a precise definition is that they are a *complete* definition. A formal treatment of *completeness* is outside the scope of this text, but two features of completeness can usefully be examined informally. The set of axioms defining the semantics of an abstract data type should be:

(i)   Complete in the sense that they define the outcome of all permissible applications of the operations of the abstract data type;

(ii)   Complete in the sense that they define operations that allow the construction of all permissible instances of the abstract data type.

The issue of completeness, in the sense of defining the outcome of all permissible operations, is examined first. It is not sufficient to note merely that the cases of, for example, **top**ping and **pop**ping an empty stack and a non-empty stack have been systematically addressed. Stack operations may be *composed,* that is, the application of an operation may be used as the source data for the application of another operation, provided such a composition is syntactically valid. Such compositions have, in fact, been used in the formulation of axioms; *axiom S1,* for example, applies **top** to the result of an application of **push**. Thus completeness, in this first sense, means that the outcome of a composition of operations (or a *stack expression,* as it is sometimes called) is also defined. For example, the outcome of

push (a, **pop**(**push** (b, **push** (c, **createstack**))))

needs to be defined, where *a, b,* and *c* are values drawn from *I.*

Exercise 2.15

What is the outcome of the stack expression

push(a, **pop**(**push**(b, **push**(c, **createstack**)))) ?

Whilst the outcome of the stack expression of the above exercise is defined, some stack expressions are undefined by the axioms of *Fig. 2.5*. For example,

top(**pop**(**createstack**))

is undefined since it involves the application of **top** (which requires a stack as source data) to the result of an application of **pop** (which, potentially, returns a result which can be a stack). However, in this particular case of **pop**ping an empty stack (from **createstack**), **pop** returns a value from M — *the stack is empty* — and the action of **top** on such a value is undefined. A similar situation arises with

pop(**pop**(**pop**(**push**(c, **createstack**))))

where **pop** is applied to a value from M. In a spirit of strictness, it could be argued that expressions such as these are not *undefined:* rather, they are *syntactically invalid*. Looking at the syntax given in *Fig. 2.5,* it can be seen, for example, that a composition of **pop**s is syntactically invalid since the result set $(S \cup M)$ is not the same as the source data set (S). Similar arguments can be made for compositions involving **top, pop,** and **push**. Such strictness is not without merit, but has the problem that many useful compositions, such as

pop (**pop** (**push** (i, s)))

are syntactically invalid, despite the fact that it is only in the specific situation where s denotes an empty stack that the invalidity arises in this case. However, clearly, the syntax of *Fig. 2.5* could be amended so that all required compositions were syntactically valid. This would not, of course, be the end of the matter, since we then need to address the real issue of **top**, **pop**, etc. being semantically undefined when applied to a value from *M*. In a sense, the argument on the syntactic validity of these undefined compositions can be construed as 'syntactic sugar' by considering the effect of adopting the suggestion, in *Section 2.3,* of incorporating the message value as part of the set *I*. If this is done, then the expressions considered above are syntactically valid when evaluated but are semantically undefined. Strictly or leniently, the semantic problem must be resolved.

The solution that we shall adopt is that of always returning the value from *M,* as the result, whenever it is used as source data for the application of an operation. There are a variety of ways of effecting this solution and ensuring that the axioms of *Fig. 2.5* are complete. New axioms can be introduced to define the result of applying **top** and **pop** to the value from *M*. For example we could have

$$\textbf{top}\,(\textit{the stack is empty}) = \textit{the stack is empty} \qquad\qquad (S7')$$
and
$$\textbf{pop}(\textit{the stack is empty}) = \textit{the stack is empty} \qquad\qquad (S8')$$

to indicate that, in these circumstances, the application of **top** and **pop** should return the source data value as the result. However, other operations are involved, such as **push** since

$$\textbf{push}\,(\text{a}, \textbf{top}\ (\textbf{pop}\ (\textbf{createstack})))$$

is also semantically undefined. In addition, it must be remembered that we are considering a specification where *M* has only one member: specifications with several different message values would complicate matters further. Adopting the approach of introducing new axioms could lead to a specification that begins to exhibit the very verbosity it is intended to avoid. A more satisfactory approach is the use of an *invariant assertion.* Appended to the axioms of *Fig. 2.5* we have the following:

invariant assertion
> *Whenever an operation is applied to a value from M then the result of the operation is that same value from M.*

This has the effect of succinctly and precisely specifying what is required. Of course, the axioms of *Fig. 2.4* are also incomplete in the sense just discussed and a similar *invariant assertion* needs to be appended.

We now conclude the examination of completeness by addressing the issue of the axioms defining operations that allow the construction of all permissible instances of the abstract data type. Intuitively, the axioms of *Fig. 2.4* are complete in this sense. Two operations — **createstack** and **push** — allow stacks to be built, and the only other operation which modifies a stack — **pop** — does not produce any stacks that could not otherwise be built. **top** and **isemptystack** do not change the contents of a stack. Hence, it is intuitive that any permissible stack can be built by a composition of **createstack**, **push**, and **pop** and that such a composition can always be expressed as a composition of **createstack** and **push** alone. A similar intuitive notion applies to *Fig. 2.5*.

Exercise 2.16
The stack expression

push(a, **pop**(**push**(b, **push**(c, **createstack**))))

constructs the stack given as the solution to *Exercise 2.15*. Write down a stack expression using only **push** and **createstack** that constructs an identical stack.

Stack expressions which involve **createstack** and **push** alone are commonly referred to as *reduced expressions*. It can be proved, as in Martin (1986), that such reduced expressions are unique, that is, there is only one reduced expression which defines a given stack. Hence, reduced expressions are said to be in *canonical form*.

2.4.4 *Summary of the axiomatic approach*

Leaving aside these digressions on messages and completeness, it is now useful to summarize the axiomatic approach. An axiomatic specification of semantics involves the implicit definition of meaning by relating the semantics of operations to one another by the use of axioms.

These axioms typically involve the definition of the result of composing operations together; a set of axioms precisely and completely defines the semantics of an abstract data type. We would not pretend that the axiomatic approach is an 'obvious' one in terms of readability, ease of production, or guidance in implementation. However, it is this very remoteness from representation that makes it an excellent device for encouraging abstraction in a specification.

In the next section, we examine the constructive approach to defining semantics and see how the benefits of axiomatic definitions can be deployed with an approach that is more accessible in terms of readability and ease of use.

Check point

(i)   What does **createstack** yield?

(ii) What is the result of
      pop(**push** b, (**push** (a, **createstack**))) ?

Solutions

(i)   A new, empty stack.

(ii)   A stack containing a single item, *a*.

2.5 Defining semantics constructively

2.5.1 Introduction to the approach

The axiomatic approach implicitly defines the semantics of operations by relating their meanings to one another. The term *implicitly* is used to emphasize the fact that, for example, **push** is not defined in isolation but always in terms of its relation to **pop, top,** and **isemptystack**. In contrast, the *constructive approach* defines the semantics of operations explicitly by relating each individual operation to the semantics of an underlying model. That is, definitions are built (or *constructed*) explicitly from operations defined in the underlying model. As we noted at the beginning of *Section 2.4,* this constructive (or *operational* or *abstract model*) approach requires that the underlying model has been (or is capable of being) precisely defined. Further details of the approach can be found in Jones (1980).

2.5.2 The underlying model of a list

The essence of the constructive approach is to select an underlying model (which can itself be regarded as an abstract data type) and to use the semantics of this underlying model to define the semantics of the abstract data type in question. Once an underlying model that is sufficiently rich and powerful has been selected and defined, it forms the basis for the direct and easy specification of a range of abstract data types using the constructive approach. A little hard work is initially done in order to make later tasks much easier than would otherwise be the case.

There is no one underlying model that is suitable for the constructive specification of all required abstract data types. Rather, an underlying model is used for the constructive specification of a range of (similar) abstract data types. In this book we shall be concerned, initially, with the abstract data types **stack, queue,** and **deque** and their variants. Such abstract data types can be broadly characterized as *linear structures,* and an underlying model of a *list* is suitable for their constructive specification. Obviously, in order to use the underlying model of a list as the basis for a constructive specification, the semantics of a list must be clearly understood (as well as precisely defined). Both clarity of comprehension and precision of definition can be achieved by examining the axiomatic specification of a list. As an aid to exposition, we begin by giving an informal description of the operations that characterize the abstract data type list. Intuitively, a list is either *empty* or consists of a *first* item and a *list* of all other items after the first, known as the *trailer*. A list is therefore an *ordered* collection.

createlist is an operation that requires no source data and produces a new, empty list as the result.

makelist is an operation that takes an item as the source data and produces a one item list as the result. For example, **makelist** applied to the item *Anne* yields the list *<Anne>*. Notice the convention used for writing down an instance of a list: the items are enumerated between a pair of angled brackets. Thus, we may have the list *alist* as

$$alist = <Margery, Ada, Fanny, Elizabeth, George>$$

concatenate is an operation that takes two lists as the source data and joins them together to produce a list as the result. For example, if the list *<Anne>* is **concatenat**ed with *alist,* we obtain the list *<Anne, Margery, Ada, Fanny, Elizabeth, George>*. Notice that **concatenate** is not commutative: *alist* **concatenat**ed with *<Anne>* yields the result *<Margery, Ada, Fanny, Elizabeth, George, Anne>*.

last is an operation that takes a list as the source data and, if the source data list is empty, produces a message indicating this as the result; otherwise it produces the last item in the list as the result. **last** applied to *alist* yields the item *George* .

leader is an operation that takes a list as the source data and produces a list consisting of all the original items, excluding the last item, as the result. **leader** applied to *alist* yields the list *<Margery, Ada, Fanny, Elizabeth>*. **leader** applied to an empty list produces an empty list. **leader** applied to a list containing only one item produces an empty list.

first is an operation that takes a list as the source data and, if the source data list is empty, produces a message indicating this as the result; otherwise it produces the first item in the list as the result. **first** applied to *alist* yields the item *Margery*.

trailer is an operation that takes a list as the source data and produces a list consisting of all the original items, excluding the first item, as the result. **trailer** applied to *alist* yields the list *<Ada, Fanny, Elizabeth, George>*. **trailer** applied to an empty list produces an empty list. **trailer** applied to a list containing only one item produces an empty list.

length is an operation that takes a list as source data and produces an integer denoting the number of items in the list as the result. **length** applied to *alist* yields the value *5*.

isemptylist is an operation that takes a list as source data and returns as the result the value *true* when the list is empty, or the value *false* when the list is not empty.

The axiomatic definition of the generic abstract data type list is now given as *Fig. 2.6,* below.

NAME
    list(item)

SETS
    L   set of lists
    I   set of items
    B   set of Boolean values consisting of *true* and *false*
    N   set of positive integers including zero
    M set of message values consisting of the single member
         the list is empty

SYNTAX
    createlist: $\rightarrow$ L
    make: I $\rightarrow$ L
    concatenate: L $\times$ L $\rightarrow$ L
    last: L $\rightarrow$ I $\cup$ M
    leader: L $\rightarrow$ L
    first: L $\rightarrow$ I $\cup$ M
    trailer: L $\rightarrow$ L
    length: L $\rightarrow$ N
    isemptylist: L $\rightarrow$ B

SEMANTICS
    $\forall i \in$ I, $\forall a, b, c \in$ L:

last(**concatenate**(a, **make**(i))) = i	*(L1)*
last(**createlist**) = *the list is empty*	*(L2)*
leader(**concatenate**(a, **make**(i))) = a	*(L3)*
leader(**createlist**) = **createlist**	*(L4)*
concatenate(**concatenate**(a, b), c) =	
concatenate(a, **concatenate**(b,c))	*(L5)*

concatenate(**createlist**, a) =

     concatenate(a, **createlist**) = a   *(L6)*

first(concatenate(**make**(i), a)) = i   *(L7)*

first(**createlist**) = *the list is empty*   *(L8)*

trailer(concatenate(**make**(i), a)) = a   *(L9)*

trailer(**createlist**) = **createlist**   *(L10)*

length(**createlist**) = 0   *(L11)*

length(concatenate(**make**(i), a)) = 1 + **length**(a)   *(L12)*

isemptylist(concatenate(a, **make**(i))) = *false*   *(L13)*

isemptylist(**createlist**) = *true*   *(L14)*

invariant assertion

> *Whenever an operation is applied to a value from M then the result of the operation is that same value from M.*

Fig. 2.6 The abstract data type **list**

The majority of the axioms can be seen to define the intended semantics in a straightforward way. For example, *axioms L1* and *L2* define the action of **last** for a non-empty list and for an empty list, respectively. This is done in a manner similar to *axioms S1* and *S6* for a stack. Considering *axiom L1,*

     concatenate(a, **make**(i))

results in a non-empty list, irrespective of whether *a* is empty or not. Hence, the result of applying **last** to the list obtained from this **concatenate** will always be an item, and will be specifically *i*.

Correspondingly, *axiom L2* defines the application of **last** on an empty list to yield *the list is empty* as its result. It does this by applying **last** to a list that will always be empty: the list obtained by applying **createlist**. A similar approach is taken with **leader, first,** and **trailer:** the action of each is defined for a non-empty list and for an empty list.

Exercise 2.17

Describe how *axioms L7* and *L8* define the application of **first** on a non-empty list and an empty list, respectively.

Axioms L11 and *L12* (listed above) merit some examination, since together they constitute a *recursive* definition of **length**. *Axiom L11* states that **length** applied to an empty list yields the result of zero. *Axiom L12* asserts that the list obtained from **concatenate**ing the list obtained from **make**(i) with a list, *a,* has a **length** one greater than the **length** of the original list, *a.* Notice that the list on the right-hand side of the axiom — the list *a* in **length**(a) — is smaller (that is, contains fewer items) than the list on the left — the list obtained by **concatenate**(make(i), a). As will be seen in the following example, repeated use of *axiom L12* will reduce the number of items in a list to which **length** is applied, correspondingly adding one to the result. Eventually, the list will become empty and *axiom L11* can then be applied. Together these two axioms define, for all particular instances of a list, what the value of its **length** will be. As the example, consider the application of **length** to the list *<Madie, Edna>*. That is,

length(<Madie, Edna>) (1)

Clearly *(1)* represents the application of **length** to a non-empty list and can therefore be written in the form of the left-hand side of *axiom L12:*

length(**concatenate** (**make**(Madie), <Edna>)) (2)

axiom L12 asserts that *(2)* is the same as:

1 + **length**(<Edna>) (3)

Repeating this process *(3)* is written as

1 + **length**(**concatenate**(make(Edna), **createlist**)) (4)

Again, *axiom L12* asserts that *(4)* is the same as:

1 + 1 + **length**(**createlist**) (5)

Using *axiom L11 (5)* can be written as:

1 + 1 + 0 (6)

which evaluates to the integer value *2*.

Exercise 2.18
Show how *axioms L11* and *L12* can be used to yield the result of
length(<Richard, Madie, Edna>).

Exercise 2.19
Complete the following

    (i)   **first(make**(i)) =   (ii) **trailer(make**(i)) =
    (iii) **leader**(**make**(i)) =   (iv) **last**(**make**(i)) =
    (v) **isemptylist**(**make**(i)) =

The solution to *Exercise 2.19* illustrates the issue of completeness, in
the sense of defining the outcome of all permissible applications of the
list operations. Put more formally, the outcome of

    leader(**make**(i))

is defined by the fourteen axioms, since *axiom L3* states that **leader**
applied to a list, *a*, **concatenate**d with the list from **make**(i), yields the
list *a*. The list *a* may be empty, in which case *axiom L3* covers the case
of the above expression.

2.5.3 Pre- and post-conditions

You will remember that our aim was to define precisely a set of
operations on a list, so that we can use these operations in a constructive
specification of an abstract data type. You can probably already
appreciate the relevance of the list operations to stack operations. For
example, **first** would seem an appropriate operation to use in the
construction of a definition of **top**. Likewise, **pop**ping a stack has
similarities with **trailer**ing a list. This similarity can usefully be
exploited is by means of *pre-conditions* and *post-conditions*.

    The principle of using pre-conditions and post-conditions can be
illustrated by means of the operation **maximum** introduced at the start
of this chapter. There, **maximum** was defined as follows:

maximum *is an operation that takes two integers as source data and returns the integer with the higher numerical value as the result (or either integer if they are equal).*

You can now appreciate that the syntax of **maximum** is

maximum: $Z \times Z \rightarrow Z$

The semantics of **maximum** are given by the following two statements.

pre-**maximum**$(x, y) ::= $ *true*
post-**maximum**$(x, y; r) ::= $ $(r >= x)$
$$\wedge \, (r >= y)$$
$$\wedge \, (r = x \vee r = y)$$

maximum has been defined by stating the relationship required between the source data and the results, without describing the computation of the latter from the former. This is achieved by specifying what must be true about the source data, by means of the pre-condition (pre-**maximum**, in this case), and by specifying what must be true about the relationship between the source data and the result, by means of the post-condition (post-**maximum**, in this case).

A *pre-condition* has the following three components.

(i)   The name of the pre-condition. It is conventional to prefix the name of the operation with *pre-* . Thus, we have pre-**maximum**.

(ii) Variables representing the source data. Conventionally, these are in lower case, in the same order as the declaration of the syntax of the operation, and are enclosed in brackets. Thus, *(x, y)* denotes two source data variables .

(iii) A declaration of the condition that must hold before the operation can be legally applied. The symbol *::=,* meaning *is defined as,* acts as a delimiter between the source data and the condition. Since **maximum** can be applied to any two integer values (that is, to any syntactically correct data) the condition is given as *true*. Examples will be given later where the condition restricts the particular values for the source data.

A *post-condition* has the following three components:

(i)   The name of the post-condition. It is conventional to prefix the name of the operation with *post-*. Thus, we have post-**maximum**.

(ii) Variables representing the source data and the result. Conventionally, these are in lower case, with source data variables given before the result variable and separated from it by a semi-colon. The order of the source data variables is the same as in the declaration of the syntax of the operation. The collection of variables is enclosed in brackets. Thus, *(x, y; r)* denotes two source data variables (*x* and *y*), followed by a result variable (*r*). Convenience is served if the symbols used for the source data variables are the same as those appearing in the pre-condition.

(iii) A declaration of the relationship that holds between source data and result after application of the operation. The symbol *::=* meaning *is defined as* acts as a delimiter between the source data and the result, and the declaration of the relationship. In this case,

$$(r >= x)$$
$$\wedge \, (r >= y)$$
$$\wedge \, (r = x \vee r = y)$$

states that the result, r, must be greater-than-or-equal-to each of the source data values x and y, and it must be equal to one or other of the source data values.

It is worth emphasizing that pre- and post-conditions do not represent applications of the operations they are defining. Rather, they represent what must be true *before* and *after* an application of an operation. Notice also how the post-condition says nothing about *how* the source data are transformed into the results. For example, where the values of the two source data variables for **maximum** are equal, nothing is said about which variable is used for assigning a value to the result. This can be contrasted with the Pascal function representing **maximum** given in *Section 2.1,* at the beginning of this chapter.

Exercise 2.20

Suppose we have a new operation **minimum**. The syntax of **minimum** is the same as that for **maximum**. Semantically, however, they are different. Informally, **minimum** takes two integers as source data and returns as the result the integer with the lower numerical value, or either integer if they are equal. Write down the pre- and post-conditions defining the semantics of **minimum**.

We now consider an example where the pre-condition is not simply defined as true. Suppose that a new operation named **restrictedmaximum** is to be defined. The syntax of **restrictedmaximum** is

$$\textbf{restrictedmaximum}: Z \times Z \rightarrow Z$$

where Z denotes the set of integers. Informally, **restrictedmaximum** is an operation that takes any two integers greater than zero as source data and returns the integer with the higher numerical value as the result. These semantics can be formally expressed in the constructive approach, by

$$\text{pre-}\textbf{restrictedmaximum}(x, y) ::= (x > 0) \wedge (y > 0)$$
$$\text{post-}\textbf{restrictedmaximum}(x, y; r) ::= (r >= x)$$
$$\wedge\, (r >= y)$$
$$\wedge\, (r = x \vee r = y)$$

Here the pre-condition states that the two integer source data values must both be greater than zero. **restrictedmaximum** is said to be *undefined* for source data values less than or equal to zero. An alternative approach is illustrated by a revised description of **restrictedmaximum**.

> **restrictedmaximum** *is an operation that takes any two integers as source data. If either of the two source data values is not greater than zero, the message* out of range *is returned as the result. Otherwise, the integer with the higher numerical value is returned as the result.*

The syntax of **restrictedmaximum** is now

> **restrictedmaximum:** $I \times I \rightarrow I \cup M$

where M denotes the set of messages containing the single member *out of range*. The semantics of **restrictedmaximum** are now

pre-**restrictedmaximum** (x, y) ::= *true*
post-**restrictedmaximum** (x, y; r) ::= *if* $(x > 0) \wedge (y > 0)$
$\qquad\qquad\qquad$ *then*
$\qquad\qquad\qquad\qquad$ $(r >= x)$
$\qquad\qquad\qquad$ $\wedge\ (r >= y)$
$\qquad\qquad\qquad$ $\wedge\ (r = x \vee r = y)$
$\qquad\qquad\qquad$ *else*
$\qquad\qquad\qquad\qquad$ r = *out of range*

Exercise 2.21

Write down the pre- and post-conditions defining the semantics of the **subtract** operation, the syntax of which was considered in *Exercise 2.7*. You should assume that integer addition (+) is the only operation that may be used in the formulation of the post-condition.

Exercise 2.22

Suppose that **subtract** is to be restricted such that it never produces a negative result, that is, it is undefined for those source data values that would produce a negative result. Write down the pre-condition that defines this.

2.5.4   The constructive specification

The underlying model of a list and the technique of pre- and post-conditions are now used to produce a constructive specification of a stack with the incorporation of messages, that is, a specification directly comparable with the axiomatic specification given in *Fig. 2.5*. Later, in

Chapter 5, other abstract data types will be specified using the constructive approach.

The *NAME, SETS,* and *SYNTAX* entries from the axiomatic approach can be used, since the constructive approach differs only in the definition of semantics. The complete specification is given in *Fig. 2.7.*

NAME
    stack (item)

SETS
    S   set of stacks
    I   set of items
    B   set consisting of the Boolean values *true* and *false*
    M   set of message values consisting of the single member
        the stack is empty

SYNTAX
    createstack: $\rightarrow$ S
    top: S $\rightarrow$ I $\cup$ M
    pop: S $\rightarrow$ S $\cup$ M
    push: I $\times$ S $\rightarrow$ S
    isemptystack: S $\rightarrow$ B

SEMANTICS
    pre-**createstack**( ) ::= *true*
    post-**createstack**(s) ::= s = **createlist**
    pre-**top**(s) ::= *true*
    post-**top**(s; r) ::= *if* s = **createlist**
                 then
                     r = *the stack is empty*
                 else
                     r = **first**(s)
    pre-**pop**(s) ::= *true*
    post-**pop**(s; r) ::= *if* s = **createlist**
                 then
                     r = *the stack is empty*
                 else
                     r = **trailer**(s)

pre-**push**(i, s) ::= *true*
post-**push**(i, s; r) ::= (**concatenate**(**make**(i), s)
pre-**isemptystack**(s) ::= *true*
post-**isemptystack**(s; b) ::= b = **isemptylist**(s)

invariant assertion
   Whenever an operation is applied to a value from M then the
   result of the operation is that same value from M.

Fig. 2.7 The constructive specification of a stack

The general style of the constructive specification is exemplified by the
entries for **top**. The pre-condition states that **top** can always be applied
to a stack. Remember that we are defining **top** so that it must produce
a message as the result when it is applied to an empty stack. This point
is achieved in the post-condition, where the value of the source data
stack, s, is used to determine whether the result, r, is an item or a
message. Where it is an item, the fact that a stack is being modelled by
means of a list is used: the item is obtained by applying **first**. Notice
the inclusion of the *invariant assertion:* it is needed so that the outcome
of all permissible stack expressions is defined.

Exercise 2.23

Re-define the *SEMANTICS* entries for **top** so that the operations
cannot be applied to an empty stack, that is, its action on an empty
stack is undefined. Write down any consequent change to the
SYNTAX entry for **top**.

Exercise 2.24

The abstract data type stack specified by *Fig. 2.7* is to be redefined so
that its size is to be restricted to no more than ten items. If an attempt
is made to increase its size beyond ten items, the message *the stack is
full* is to be the result of the appropriate operation. Write down the
changes to the specification needed to incorporate this change.

In contrast, it is comparatively difficult to incorporate such size
constraints into an axiomatic specification. For example, to carry out a

version of *Exercise 2.24* for the corresponding axiomatic specification would require the inclusion of axioms defining a **length** operation, which could then be included in the axioms defining **push**. This **length** operation would have a peculiar status: it would be a necessary part of the axiomatic definition, yet it would not be an operation that was directly available to a stack user. Such operations are said to be *hidden*. The issue of *hidden* operations will be examined again in *Chapter 7*.

2.5.5 Summary of the approach

In contrast to the axiomatic approach, the constructive approach defines the semantics of operations explicitly by relating each individual operation to the semantics of an underlying model. As you have seen, the operations may be related to the underlying model by pre- and post-conditions. Constructive specifications are easier to read and to write than axiomatic specifications — at least for people whose motivation is principally to produce specifications that will help in the task of devising an appropriate representation. However, this ease of use requires that the underlying model must be capable of precise definition. The most useful way of employing the two approaches is to define an underlying model sufficiently powerful for it to be used in the constructive specification of a range of abstract data types. *Exercise 2.25*, below, demonstrates this utility by considering the specification of a new abstract data type, albeit a variant of a stack. The use of the approach in the specification of other abstract data types will be considered in later chapters.

Exercise 2.25
A new abstract data type, *otherstack*, is to be specified using the constructive approach. The operations that define *otherstack* include all those given in the specification of a stack in *Fig. 2.7*. In addition, there is a further operation: **bottom**. **bottom** is an operation that takes an *otherstack* as source data and, if the *otherstack* is empty, produces the message *the otherstack is empty* as the result; otherwise it produces as the result the item at the bottom of the *otherstack* (but does not delete it). Using *Fig 2.7* as a guide, write down a constructive specification of the abstract data type *otherstack*.

Check point

(i)   Why is **makelist** a necessary operation in the underlying model of a list?

(ii) What is the purpose of a *pre-condition?*

(iii) What is the purpose of a *post-condition?*

Solutions

(i)   There is no operation that directly adds an item to a list; only the **concatenate** operation that adds lists together. Hence **makelist** converts an item into a list so that it may then be **concatenate**d with another list.

(ii) A pre-condition asserts what must be true about the source data *before* an operation is applied.

(iii) A post-condition asserts what must be true about the relationship between the source data and the result *after* an operation has been applied.

3 Representation and implementation of abstract data types

3.1 Introduction

Our aim in this chapter is to show a number of different ways of implementing a stack. We shall develop each implementation from the formal specification given in *Chapter 2*. Our purpose is two-fold:

(i) to investigate how to transform a formal specification into an implementation;

(ii) to discover what features are desirable in a programming language to support the process of implementation.

Going from a specification to an implementation is a two-stage process. First of all, we investigate the data structures that will be used to *represent* the abstract data type within a program. For example, we could choose to represent the abstract data type **employeestack** as an array of string values. Each string would then represent an employee (possibly containing either the employee's name or, perhaps, a staff code), and the operations of pushing and popping would be equivalent to adding or deleting items from the array. Once the representation has been decided upon we embark on the *implementation* stage, in which the chosen data structures and their associated operations are turned into program code.

We have separated out the design of the data structures from the actions of writing a program because it is quite conceivable that we would like to use a data structure which is not offered by our chosen language. For example, we might choose to represent a stack using links — to show the order of items in the stack — but not have pointers

53

available in our selected language. In this case an implementation using an array might be possible in which the links are implemented as index values. No programming language offers every data structure that a programmer might require. Pascal, for example, does not directly support stacks but, being a general-purpose language, it does offer sufficient data structures for the construction of a stack data structure in a very straightforward way.

We begin our investigations by looking at implementations in Pascal. As will always be the case, we aim for an implementation that exactly meets the requirements of the specification. That is, we aim for an implementation that precisely supports the five stack operations (**createstack, top, pop, push,** and **isemptystack**), and *nothing more or nothing less*.

The implementation of a stack in Pascal will reveal certain inadequacies in our approach and will highlight deficiencies in the Pascal language for implementing abstract data types. You will see that the concept of information hiding is central to good implementation. These investigations will lead us to a set of requirements for an ideal programming language. It turns out that Standard Pascal offers very little in this respect. Fortunately, there have been some important developments in the design of high-level languages which directly address this problem. MODULA-2 and Ada are examples of languages specifically designed to support information hiding and we shall examine them in the next chapter.

Exercise 3.1
Distinguish between the terms: *specification, representation,* and *implementation* when applied to an abstract data type.

3.2 A cursor-based implementation of a stack in Pascal

The basic idea is to represent a stack by two objects: an array (to hold the data items) and a cursor (to indicate the position, in the array, of the top-of-stack item). As items are pushed on and popped off the stack, the top item changes and the value of the cursor reflects this fact. Typically the cursor will provide the index of the top item.

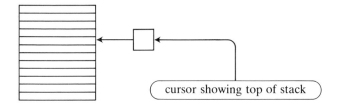

cursor showing top of stack

Clearly, an array representation restricts the total number of items that can be held on the stack at any time.

Pushing an item onto the stack is achieved by placing the data into an unoccupied element of the array and making the cursor refer to that element. To be successful, the pop operation needs information about the sequence in which items were pushed onto the stack. There are many ways of recording this information, and we shall employ one of the simplest. In our method items on the stack will be made to occupy contiguous elements of the array located at one end of the array.

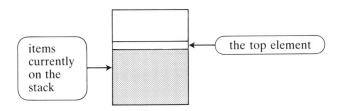

items currently on the stack

the top element

Pushing an item onto the stack will result in the data being added into the free element of the array next to the current top of stack. Popping the stack simply means setting the cursor to point to the array element immediately before the current top of stack element.

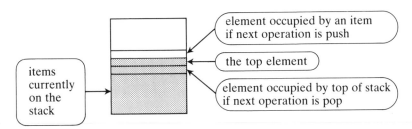

element occupied by an item if next operation is push

the top element

items currently on the stack

element occupied by top of stack if next operation is pop

Thus, the sequence of items in the stack is reflected in the sequence of occupied elements in the array, and the cursor moves up and down the array to show the position of the top-of-stack element.

The next step is to consider the implementation in Pascal. We shall choose an implementation in which the bottom of the stack is firmly anchored to the low-indexed end of the array. The stack is then allowed to grow towards the top end of the array, i.e. towards the high-indexed end. The current position of the top of the stack is held as a cursor — a variable, named *Top,* holding the index position of the top item of the stack. The situation is as shown in *Fig. 3.1,* in which the array has index values from *1* to *MaxSize.*

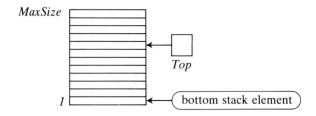

Fig. 3.1   An array representation of a stack

This data structure can be implemented in Pascal with the declarations listed below.

```
const
  MaxSize = 100;
type
  ItemType = {whatever is required};
  Cursor = 0 .. MaxSize;
  Storage = array[ 1 .. MaxSize ] of ItemType;
  Stack = record
            Top: Cursor;
            Item: Storage
          end;
```

ItemType will eventually be defined to be a type that meets the needs of the particular stack application. The type *Cursor* is a subrange of the integers which includes zero. This is to enable us to signify when the stack is empty.

Stacks can now be defined by such declarations as are listed below.

var
> *FirstStack, SecondStack, ThirdStack: Stack;*

These stacks are implemented by three separate records, each consisting of two fields. The first field holds the cursor for the top of the stack, and the second holds the array itself. You should note that the cursor and the array are aggregated into a single structure by means of a record. This means that the complete implementation of a stack can be referred to by a single name — *Stack*. Thus to access the top item of the stack named *SecondStack* you would write

> *SecondStack.Item* [*SecondStack.Top*]

Exercise 3.2
Using the definitions above, write down the Pascal code which:

(i)   declares a stack called *NameStack*;

(ii) accesses the item immediately below the top item;

(iii) shows how the condition that *NameStack* is empty is denoted.

The fact that an array, a finite structure, is being used to represent an infinite structure (a stack is a form of sequence), means that the implementation is limited in the number of items that can be on the stack at any time. In the above implementation the constant *MaxSize* defines the maximum size of a stack. This was not something that we had to consider in the formal specification. Thus, an implementation can contain features which have been introduced solely from the representation and which have nothing to do with the specification. The separation of specification from representation and implementation makes us very aware of the decisions that have to be taken in the design process.

The five operations that were formally specified in *Chapter 2* can now be implemented in Pascal. We shall write one procedure (or function, as appropriate) for each operation. We shall begin with **createstack** whose purpose is to initialize an empty stack by setting the cursor to zero.

```
procedure CreateStack ( var S: Stack);
begin
   S.Top := 0
end {CreateStack};
```

A typical call of this procedure is

```
CreateStack (FirstStack);
```

where the variable *FirstStack* must have been declared previously to be of type *Stack*. Note that, prior to this call, the value of the variable *FirstStack* is undefined and any attempt to use it, say by invoking procedures for pushing and popping, could end in disaster. It is important that anyone wanting to use these stack manipulation procedures remembers to invoke *CreateStack* before invoking any of the other procedures. After the call to *CreateStack,* the stack is initialized but is empty.

You will have noticed that the operation **createstack** returns a single value as its result (a stack). This operation is therefore a candidate for implementation as a function in Pascal. Unfortunately, structured objects (a stack is being implemented as a record type) cannot be returned by Pascal functions, and we are forced to use a procedure instead. Clearly, the parameter to *CreateStack* must be a **var** parameter since a stack is required as its result. Strictly speaking, a **var** parameter does not accurately implement the specification since it permits data to be input into the procedure. However, this is of little consequence here because our implementation of the procedure does not expect data to be passed into it.

The operation **isemptystack** can, however, be implemented as a function.

```
function IsEmptyStack (S : Stack): Boolean ;
begin
   IsEmptyStack :=  (S.Top = 0)
end {IsEmptyStack};
```

The operation **top** also has only one result — the value of the item at the top of the stack. Whether we can implement it as a function will depend on the type of the stack items to be manipulated. In general, therefore, we have to implement **top** as a procedure.

```
procedure Top (S: Stack;  var Item: ItemType);
begin
  if IsEmptyStack (S)
  then
    Writeln (' The stack is empty.')
  else
    Item := S.Item [ S.Top ]
end {Top};
```

When an attempt is made to access the top item in an empty stack we have chosen simply to write out a message at the user's terminal. This is not strictly in accordance with the specification. The specification requires that a value is returned but that this value is *either* a stack item *or* a message.

Whatever situation is detected by **top** is always communicated to the operation that invokes it. This is not the case with our implementation. The difficulty in Pascal with its strong typing is in implementing the union of sets. There are several ways of overcoming this difficulty which we shall discuss in *Section 3.5*.

Here is our implementation of **pop**:

```
procedure Pop (var S: Stack);
begin
  if IsEmptyStack (S)
  then
    Writeln (' The stack is empty.')
  else
    S.Top := S.Top - 1
end {Pop};
```

In essence, all that happens here is that the cursor is reduced by one when the stack is non-empty. But this is significant in another respect. The specification says that a *new* stack is to be produced as the result of this operation. That is, there are two different stacks involved in the operation — the stack which is input to the operation and the resultant stack produced by the operation. Our implementation involves only one instance of a stack. A single stack is *updated* by our procedure *Pop* — hence the use of a single **var** parameter. Our implementation saves storage space; for a programmer to retain a copy of the stack as it exists prior to executing the *Pop* procedure would require making an explicit copy before invoking *Pop*.

The final operation to be implemented is **push**.

```
procedure Push (Item: ItemType; var S: Stack);
begin
  if S.Top = MaxSize
  then
    Writeln (' The stack is full.')
  else
    begin
      S.Top := S.Top + 1;
      S.Item [ S.Top ] := Item
    end
end {Push};
```

Exercise 3.3

A distinction was made in *Chapter 2* between two types of constraints: *inherent* and *representation-dependent*. To which types do the three messages in *Top, Pop,* and *Push* belong?

3.3 A pointer-based implementation of a stack In Pascal

Here we shall represent a stack as a group of linked records. Each record will represent one item on the stack. In addition there is a link to the top item of the stack. *Fig. 3.2* depicts this representation.

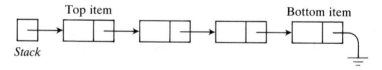

Fig. 3.2 A linked representation of a stack

This representation views a stack as a link to the record of its top item. All other items can be accessed through the links from one record to the next. An empty stack will be represented by the link *Stack* having the value **nil** (i.e. not referring to any record).

The definitions required to implement this linked representation are:

type
   ItemType = *{whatever is required for the application };*
   Link = ↑ *StackRecord;*
   StackRecord = **record**
                Item: ItemType;
                Previous: Link
        end;
   Stack = *Link;*

At first sight you may think it unnecessary to have the last definition in this list, *Stack = Link,* because it simply introduces a new identifier for a type that already exists. You would be wrong to think this. An implementation must always have a type identifier of the same name as the abstract data type. This will ensure that a programmer can specify instances of the abstract data type by such declarations as:

var
   FirstStack, SecondStack, ThirdStack: Stack;

In this way the programmer can create and manipulate instances of the abstract data type as though it were directly available in Pascal.

Since we are using pointers we shall have to remember to allocate some new storage space for each new item added to the stack. This, of course, is achieved with the Pascal procedure *new.*

Exercise 3.4
Implement the operation **createstack** using the above pointer-based definitions.

Exercise 3.5
Implement the operation **isemptystack** as a function.

Next we look at the implementation of the operation **top**. Its basic structure is similar to the corresponding cursor-based implementation.

```
procedure Top (S: Stack; var I: ItemType);
begin
  if IsEmptyStack (S)
  then
    Writeln ('The stack is empty ')
  else
    I := S↑.Item
end {Top};
```

Exercise 3.6
Implement the operation **pop**. Also, dispose of the storage space released when the top item from the stack has been removed.

Finally, **push** is implemented as

```
procedure Push (Item: ItemType; var S: Stack);
var
  Temp: Link;
begin
  new (Temp);
  Temp↑.Item := Item;
  Temp↑.Previous := S;
  S := Temp
end {Push};
```

Exercise 3.7
Are there any messages in the above pointer-based implementation which arise from representation-dependent situations?

A word of caution: it is often tempting to use Pascal assignment as a way of copying a value, but this can lead to unexpected results. To see why, let *S1* and *S2* be variables of type *Stack* and suppose that the assignment *S1* := *S2* is made. Our stack implementation means that *S1* and *S2* would point to the same stack items and, therefore, any operation performed on either *S1* or *S2* will result in both stacks being modified, which may not be the intended result. It would be sensible to provide an additional stack operation, *Copy,* say, to perform the necessary copying of all the items in a stack. Similarly, it is also a good idea to provide a print routine as a matter of course for any abstract data type implementation. It is surprising how often a user wants to see the current contents of an abstract data type!

3.3 Information hiding

We have presented two quite different implementations, in Pascal, for the abstract data type **stack**. These are just two of many possible implementations, and precisely which one to choose often depends on the application. What is important, however, is that you have been presented with *different* implementations for the *same* specification, and both implementations conform to the syntax and semantics of the formal specification. Since the two implementations are so different it is instructive to compare them.

In both implementations we defined an appropriate data structure and defined a type whose identifier was the name of the abstract data type, *Stack*. Also, both implementations had a separate procedure or function for each of the five stack operations.

A representation-dependent constraint concerned with a full stack was introduced in the cursor-based implementation. No such constraint was introduced with the linked representation.

In practice, the cursor-based implementation can turn out to be wasteful of computer storage because the array will not often be full, yet storage has to be set aside for the whole array. The pointer-based implementation uses precisely the amount of storage required to hold the items currently on the stack. However, it does require additional storage to hold the pointers, but this is seldom a heavy overhead.

The most significant fact, from the point of view of a programmer wishing to *use* a stack implementation, is that the headings of the five procedures and functions are *identical* in both implementations. For example, the heading for the procedure *Push* is:

procedure *Push(Item: ItemType;* **var** *S: Stack);*

in *both* implementations. Notice that it is impossible to tell, from this heading alone, which implementation is being used. The same is true of the other procedures and functions. This is an example of *information hiding*. The details of the specific implementation are hidden inside the bodies of the procedures and functions. Here, unfortunately, these details are not very well hidden since you can see them just by looking at listings of the procedures and functions. What we would like to be able to do is to make the details totally concealed and inaccessible.

To understand why information hiding is important you must first appreciate that, in general, there are two separate groups of people who are interested in the abstract data types. First, there are the people who implement the abstract data types and who we shall refer to as the *implementors*. The other group, known as application programmers, are *users* of an implementation. You can think of the implementors as suppliers of the software which implements an abstract data type, and the application programmers as consumers of that software.

A useful analogy is the supply and use of a washing machine. The manufacturer of a washing machine is the supplier, whereas the consumer is the person who uses the machine for washing. The consumer would not usually know, or be expected to know, the details of how the machine actually works. The consumer is more concerned with the machine's functioning — what operations it can perform and how to make it do the necessary washing. The supplier does not normally expect the consumer to delve into the innards of a machine either to mend it or to change its functions. If a machine breaks down or fails to perform some operation, it will be repaired by the manufacturer or some other expert. This is precisely the situation that we would like to be in with software for abstract data types. The application programmer should be interested only in the functioning of a piece of software. Any maintenance of the software which requires a knowledge of the details of implementation is the province of the implementor.

Our primary motive for pursuing this point of view is that, by separating the implementation of an abstract data type from its use:

(i)   an implementation can be changed without knock-on effects to the application program. When a representation or an implementation has to be changed (on the grounds of efficiency, say), this can be achieved without affecting the application *in any way*;

(ii) Maintenance of an implementation can take place without affecting the application, and *vice versa*.

The role of the formal specification is, of course, to unambiguously define *what* is required. The specification determines *what* the implementor should provide and tells the application programmer *what* to expect.

The task of building a piece of software for use by others therefore requires programming language support to enable the supplier to hide the details of the implementation from the customer. The first step in this direction is the use of good programming practices. You have just seen how it is possible to confine the details of an implementation to the bodies of a set of routines. Also, by sensible design, the user of those routines cannot tell from the parameters of the routines what representation or implementation has been used. To be able to use a routine the application programmer needs only the following information:

(i)   the name of the routine;
(ii) information about its parameters;
(iii) a description of the purpose of the routine.

Provided that care is taken in the design of the parameters, by which we mean hiding the underlying representation by using suitable type definitions, Pascal enables us to offer a certain degree of information hiding. Pascal's drawback is that, in order to compile an application program, all procedures, functions, and type definitions have to be present. The result is that all the details of the implementation are visible to the application programmer. This is a serious situation because:

(i) by looking at the implementation, an application programmer might be unduly influenced by what is there, and write the application in such a way that it becomes dependent on the implementation of the abstract data type. For example, if the applications program is required to determine the number of items currently on a stack, and the cursor-based implementation of *Section 3.2* is used, the programmer might be tempted to use the current value of the cursor to determine the number on the stack. In this situation, changes to the implementation will require changes to the application. The implementation and the application are no longer independent.

(ii) the application programmer would be in a position to alter the implementation or include new operations. In either case the implementation would no longer match the original specification. For example, an application might always pop the stack having ascertained what the top item is. Here, the efficiency of the stack operations can be improved if the *Pop* procedure were so altered that the top element were also returned as a result, thus avoiding having to invoke both *Top* and *Pop*. In such a case, the *Top* procedure might be withdrawn from the 'revised' implementation because it was no longer being used. Another application programmer wanting to use a stack might use the revised implementation unknowingly and spend may fruitless hours trying to find   bugs in the software on the assumption that the implementation is faithful to the specification.

(iii) whenever you have the situation in which there exist multiple copies of a piece of software there is always the danger of inconsistency occurring when the software is updated. It is usually impossible to update all copies at the same time (some people may never discover that their copy is out of date). To avoid the situation where application programs behave differently when executed with different versions of an abstract data type, it is imperative that all versions conform to the same specification. Thus, the updating of software can take place over a period of time without disrupting existing applications.

In conclusion, we would like to ensure that an implementation of an abstract data type is *re-usable* (i.e. it can be used by different applications), *independent* (of any application) and *maintainable* (avoids the multiple update problem).   Therefore, if we use programming languages which support information hiding we shall be in a strong position to ensure cost-effective and easily maintainable software.

Exercise 3.8
Write down those parts of a stack implementation that an *application programmer* needs to know about in order to construct an application that uses a stack.

Exercise 3.9
Describe briefly the benefits to be gained by implementing an abstract data type as a set of procedures and functions in which each abstract data type operation is implemented by a separate routine.

Exercise 3.10
What is the *purpose* of information hiding?

Check point

(i)   How should the operations which specify an abstract data type be implemented?

(ii) What are the guidelines for producing parameters for the procedures/functions that implement abstract data type operations?

(iii) What is information hiding?

Solutions

(i)   One procedure or function is written for each operation. As far as possible the name of the procedure/function should correspond to the name of the operation.

(ii) The parameters should correspond to the syntax of the operation and should not reveal anything about the implementation. There should be a parameter with the same name as the abstract data type.

(iii) It is separating the method of implementation from the procedure headings in such a way that the implementation is hidden from the user (application programmer) of the procedures.

3.4 Implementing a stack in UCSD Pascal

Ideally we should separate the information about the implementation which the application programmer needs from the remaining implementation details. That is, the type definitions (of the procedures and functions and their parameters) should be separated out. UCSD (University of California, San Diago) Pascal supports a construct called a **unit** which aids this separation. A UCSD Pascal unit has the following structure:

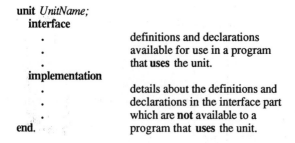

```
unit UnitName;
   interface
        .                   definitions and declarations
        .                   available for use in a program
        .                   that uses the unit.
   implementation
        .                   details about the definitions and
        .                   declarations in the interface part
        .                   which are not available to a
   end.                     program that uses the unit.
```

Fig. 3.3   The structure of a UCSD Pascal unit.

A **unit** is a *separately compiled* piece of program which can be used (or shared) by any other piece of program that includes an appropriate **uses** statement. For example, here is the structure of a main program that makes use of a unit:

```
program UnitUser;
uses UnitName;
     .                   further definitions and declarations
     .                   needed by the main program.
begin
     .                   body of the
     .                   main program.

end {UnitUser}.
```

As a first step towards understanding how a unit works, imagine a complete Pascal program with many definitions and declarations of constants, types, variables, procedures, and functions. Some of these definitions and declarations are concerned with the implementation of the **stack** abstract data type and are bodily removed into the **interface** part of a unit. When the main program is compiled, the compiler obtains the definitions and declarations from the interface part of the named unit and combines them with the definitions and declarations remaining in the main program, a process which you can think of as reproducing the original monolithic program.

The way that units work in UCSD Pascal means that any **uses** statement must be placed *immediately* after the program heading, and it is likely that the **uses** statement will have to have a *compiler directive* to specify which file the unit is stored in. At the time the main program is compiled, the effect is as if the **uses** statement is replaced by the contents of the *interface* part of the named unit. It is possible to name more than one unit in a **uses** statement.

As an example of how to use UCSD units we shall first look at the structure of a standard monolithic Pascal application program that utilizes a stack (but does not use *units*). The application reads in a sequence of integers, stores them on a stack, and then writes out the numbers in the reverse order from which they were read in. The sequence is terminated by the value –*999*. Here we shall employ the cursor-based implementation of a stack.

```pascal
program Reversal;
const
  MaxSize = 100;
  EndOfListMarker = -999;
type
  ItemType = Integer;
  Cursor = 0 .. MaxSize;
  Storage = array[ 1 .. MaxSize ] of ItemType;
  Stack = record
             Top: Cursor;
             Item: Storage
           end;
var
  NumberStack: Stack ;
  CurrentNumber: ItemType;

procedure CreateStack ( var S: Stack);
begin
  S.Top := 0
end {CreateStack};

function IsEmptyStack (S: Stack): Boolean;
begin
  IsEmptyStack := (S.Top = 0)
end {IsEmptyStack};

procedure Top (S: Stack; var Item: ItemType );
begin
  if IsEmptyStack (S)
  then
    Writeln (' The stack is empty.')
  else
    Item := S.Item [ S.Top ]
end {Top};
```

```
procedure Pop (var S: Stack);
begin
  if IsEmptyStack (S)
  then
    Writeln ('The stack is empty.')
  else
    S.Top := S.Top - 1
end {Top};

procedure Push (Item: ItemType;  var S: Stack) ;
begin
  if S.Top = MaxSize
  then
    Writeln (' The stack is full.')
  else
    begin
      S.Top := S.Top + 1;
      S.Item [ S.Top ] := Item
    end
end {Push};

begin                              {Start of main program}
  CreateStack (NumberStack);
  Write ('Enter first number: ');
  Readln (CurrentNumber);
  while CurrentNumber <> EndOfListMarker do
    begin
      Push (CurrentNumber, NumberStack);
      Write ('Enter next number: ');
      Readln (CurrentNumber)
    end;
  Writeln (' The numbers in reverse order are:');
  while not IsEmptyStack (NumberStack) do
    begin
      Top (NumberStack , CurrentNumber);
      Writeln (CurrentNumber);
      Pop (NumberStack)
    end
end {Reversal }.
```

Not only is the simple program quite lengthy it is also difficult to distinguish details of the application from details of the implementation of the stack. The use of a unit will clarify the situation. We shall remove all the stack operations together with all attendant constant and type definitions into a unit named *IntegerStackOps*. In so doing, we shall place the bodies of the routines into the **implementation** part of the unit.

```pascal
unit IntegerStackOps;
  interface
    const
      MaxSize = 100;
    type
      ItemType = Integer;
      Cursor = 0 .. MaxSize;
      Storage = array[ 1 .. MaxSize ] of ItemType;
      Stack =    record
                    Top: Cursor;
                    Item: Storage
                 end;

    procedure CreateStack ( var S: Stack);
    function IsEmptyStack (S: Stack): Boolean;
    procedure Top (S: Stack; var Item: ItemType);
    procedure Pop  (var S: Stack);
    procedure Push (Item: ItemType; var S: Stack);

implementation
    procedure CreateStack;
    begin
      S.Top := 0
    end {CreateStack};

    function IsEmptyStack;
    begin
      IsEmptyStack := (S.Top = 0)
    end {IsEmptyStack};

    procedure Top;
    begin
      if IsEmptyStack (S)
      then
        Writeln (' The stack is empty.')
      else
        Item := S.Item [ S.Top ]
    end {Top};

    procedure Pop;
    begin
      if IsEmptyStack (S)
      then
        Writeln (' The stack is empty ')
      else
        S.Top := S.Top – 1
    end {Pop};
```

```
procedure Push;
begin
  if S.Top = MaxSize
  then
    Writeln (' The stack is full.')
  else
    begin
      S.Top := S.Top + 1;
      S.Item [ S.Top ] := Item
    end
end {Push};
end. {of IntegerStackOps}.
```

The name of the unit, *IntegerStackOps,* reflects its purpose: it contains the operations for a stack of integer items. The construction of the unit mirrors what we want from information hiding. The **interface** part contains the information that the application programmer requires — the headings of the routines — whereas the **implementation** part contains the bodies of the routines. In the **implementation** part the bodies of the routines have headings which *do not contain the formal parameters.* These headings serve merely to associate the bodies with their corresponding full headings in the **interface** part.

The corresponding application program now looks like this:

```
program Reversal;
uses StackOps;        {This may need additional information to }
                      {identify the file in which the unit is stored; }
                      {it depends on version of the UCSD system.  }

const
  EndOfListMarker  = –999;

var
  NumberStack: Stack;          {using type definitions held }
  CurrentNumber: ItemType;     {in the unit.}

begin                          {Start of main program}
  CreateStack (NumberStack);
  Write ('Enter first number: ');
  Readln (CurrentNumber);
  while CurrentNumber <> EndOfListMarker do
    begin
      Push (CurrentNumber, NumberStack);
      Write ('Enter next number: ');
      Readln (CurrentNumber)
    end;
  Writeln (' The numbers in reverse order are:');
```

```
    while not IsEmptyStack (NumberStack)  do
       begin
          Top (NumberStack, CurrentNumber);
          Writeln (CurrentNumber);
          Pop (NumberStack)
       end
    end {Reversal}.
```

The main program now concentrates on the needs of the application, independently of the implementation of the stack.

Exercise 3.11

What changes would have to be made to the new version of *Reversal* if the pointer-based implementation of the stack were to be used?

We can be critical of the unit *StackOps* in several ways. But first, the good points. Clearly, the main program is much improved, both in size and readability. Also, since UCSD Pascal prevents any information contained within the implementation part of a unit being used outside the unit, it is only the information contained within the interface that can be accessed by a program that **uses** the unit.

There are two major disadvantages of this scheme, which are well illustrated in the above unit.

(i)   The whole of the data structure that represents the stack (i.e. the array and the cursor) resides in the *interface* part of the unit and, therefore, is available for use within the application program. The data structure is *not* hidden in the way that we would like.

(ii)   The type of the stack items is declared in the unit. This means that the above unit defines a stack of integer items. If we write an application program to manipulate a stack of characters, say, then a completely separate unit would be required.

Both problems arise from Pascal's declare-before-use rule and the fact that units are compiled separately from the main program. To be allowed to include routine headings within an interface part means that all necessary type definitions for the parameters must also appear in the interface part — prior to the headings. If this were not done, the unit could not be compiled in the usual Pascal way since the formal parameters would be reached before their types had been defined.

There is a partial solution to the problem of holding the definition of the data structure in the interface part of a unit. We can arrange to have the data structure defined in the implementation part, provided there is a corresponding change to the headings of the routines, as follows:

```
unit IntegerStackOps;
   interface
      type
         ItemType = Integer;
      procedure CreateStack;
      function IsEmptyStack: Boolean;
      procedure Top (var Item: ItemType);
      procedure Pop;
      procedure Push (Item: ItemType);

   implementation
      const
         MaxSize = 100;
      type
         Cursor = 0 .. MaxSize;
         Storage = array[ 1 .. MaxSize ] of ItemType;
         Stack = record
                     Top: Cursor;
                     Item: Storage
                  end;
      var
         S: Stack;                    { This is new }

      procedure CreateStack;
      begin
         S.Top := 0
      end {CreateStack};

      function IsEmptyStack;
      begin
         IsEmptyStack := (S.Top = 0)
      end {IsEmptyStack};

      procedure Top;
      begin
         if IsEmptyStack (S)
         then
            Writeln ('The stack is empty.')
         else
            Item := S.Item [ S.Top ]
      end {Top};
```

```
    procedure Pop;
    begin
      if  IsEmptyStack (S)
      then
          Writeln (' The stack is empty .')
      else
          S.Top := S.Top – 1
    end {Pop};

    procedure Push;
      begin
      if  S.Top = MaxSize
      then
          Writeln (' The stack is full')
      else
          begin
            S.Top := S.Top + 1;
            S.Item [ S.Top ] := Item
          end
    end {Push};
  end. {of StackOps}.
```

The data structure is now hidden within the implementation part. However, because the type definitions are only available for use in the implementation part of the unit, the declaration of the stack itself (the **var** declaration) also has to be within the implementation part. This means that the variable S cannot be accessed within the application program and, therefore, cannot be referred to in the parameters of the routines. Effectively, what was previously a formal parameter, S, has had to become global to all the routines. Since the **uses** statement occurs in the main program, the global variable, S, exists for the whole of the execution of the program. Specifically, this means that S retains its values between calls of the stack routines. The type of the stack items, *ItemType*, still has to be declared in the interface part because it is required by both the application program and the routines in the implementation part. (Any object defined in the interface part of a unit is accessible both in the implementation part and in the program that uses the unit.)

This method of implementing the stack abstract data type is not to be recommended for three main reasons.

(i) Only one physical stack can exist in each unit. An application program cannot manipulate more than one stack without having a separate unit for each instance of a stack, and to do this means having different names for the routines in each unit.

(ii) The fact that the stack data structure is omitted from the parameter lists of the routines makes the purpose of each routine less clear to an application programmer.

(iii) An application program becomes more difficult to maintain because it is not clear from looking at the invocations of the stack routines exactly what is being modified.

The use of units in UCSD Pascal helps with information hiding in so far as it does separate out the implementation of an abstract data type from its use in an application program. The degree of information hiding is, however, small because, whatever technique is used, both the interface and implementation parts of a unit are inseparable and must be together when it is compiled. The inseparability means that the implementation details are visible and, unless the data structure is placed in the implementation part, the details of the implementation can still be used by the application programmer. However, placing the actual data structure in the implementation part is not an acceptable strategy because it hides the very information that the application programmer needs.

A final disadvantage of UCSD units concerns the position of the definition of the type *ItemType*. Since the definition of *ItemType* is held within a unit, the routines within that unit can manipulate items of that type alone. Therefore, if an application program needs to operate on stacks with differing types of item, separate and complete units are demanded for each instance of a stack needing a different *ItemType*.

Exercise 3.12
List the advantages and disadvantages of UCSD Pascal units for the purpose of information hiding.

This discussion has concentrated on the advantages and disadvantages of UCSD units. However, it has also focused attention on the kinds of facility that we would like to have in our ideal mechanism for information hiding. Information hiding is achieved through what are called the **encapsulation** facilities in a programming language. Based on what we have just seen, we can list the capabilities we would ideally like to have in a programming language.

(i) To separate the whole of the implementation of an abstract data type from an application program;

(ii) To limit access to the implementation of an abstract data type to the headings of the routines which implement the operations;

(iii) To allow the application program to manipulate more than one instance of the abstract data type without having to replicate the implementation;

(iv) To separately compile the implementation of an abstract data type and make the implementation easily available to more than one application program;

(v) To be able to change the implementation of the abstract data type without in any way affecting an application program;

(vi) To be able to define the type of the abstract data type item in the application program.

The last point recognizes the fact that, in specifications, we have the ability to define *generic* abstract data types, and would like the same facility in programming languages.

The above list is *ideal,* it does not imply that all these requirements are, or can be, met in a single language.

3.5   Constraint handling

So far in our discussion of the implementation of abstract data types we have avoided any direct reference to issues relating to constraints. The majority of abstract data types include constraints. For example, one stack constraint is that an empty stack cannot be **top**ped. Other constraints come from the application, for example, the desire to restrict the maximum number of items in a stack. The methods of specification that we have presented return a message value whenever an attempt is made to violate a constraint. The problem, from an implementation point of view, is that an operation is permitted to return values of different types, a difficult thing to achieve with strongly typed languages like Pascal.

Many texts on abstract data types view such constraints as not being able to top an empty stack as errors. We have studiously avoided using the term 'error', particularly when discussing specifications. The reason is simple: an error is, literally speaking, a mistake. In our specifications we have always taken constraints into full account; there is no mistake here. Some would argue that, in an application, if one of our messages is returned, it signifies that a mistake has been made and the implementation should take appropriate action. The counter argument is that mistakes in an application program are the province of the application programmer and not of the specifier or implementor of the abstract data types. The nature of such an error depends solely on the application and there is no satisfactory way that an implementation can hope to deal with all such mistakes.

In our view the specification of an abstract data type must explicitly recognize the existence of constraints. Any implementation must, therefore, reflect this recognition by incorporating suitable tests and providing some method of reporting that such an event has occurred. However, constraint violations occur because of problems within the application program, so it is up to the application program to deal with such situations. There are two strategies for dealing with constraint violations in an application program.

(i) Detect a constraint violation *before* it arises, and take evasive action.

(ii) Allow a constraint violation to occur, let the implementation of the abstract data type detect the situation, and take remedial action *after* the event.

The former strategy can be implemented by ensuring that each invocation of an abstract data type operation is *guarded*. That is, a test is made, by the application program, to see whether a proposed operation would lead to a constraint violation being detected *before* invoking the relevant routine. For example, the stack routine *Top* would always be be called within a construct similar to the following:

```
if IsEmptyStack (TheStack)
then
    { take appropriate action }
else
    Top (TheStack)
```

This strategy forces the application programmer always to consider what action the application is to take in these situations.

The implementations we have described so far are adequate to support this strategy because no special action needs to be taken by the implementation to deal with constraint violations. If a constraint violation is detected by our implementation, it signifies a mistake in the application program and a simple message is all that is required (we might then abort processing when this happens — if the programming language allows it).

The alternative strategy means that the implementation has to detect constraint violations *and report back to the application program*. It is up to the application program to take appropriate action *after* the abnormal event has occurred. For example, look once again at our cursor-based implementation of the stack operation **top**:

```
procedure Top (S: Stack;  var Item: ItemType);
begin
    if IsEmptyStack (S)
    then
        Writeln (' The stack is empty.')
    else
        Item := S.Item [ S.Top ]
end;
```

In the situation when the stack is empty the error message '*The stack is empty.* results, but no value is returned via the parameter, *Item*. Since the program that invoked *Top* will, in general, be expecting an item to be returned — that's why *Top* was called in the first place — erroneous results might ensue. The calling program will not know that *Top* has detected a constraint violation. Clearly, we must pass information from the procedure to its calling program indicating that such an event has occurred and leave it to the calling program to take appropriate action. There are two approaches to this problem.

(i) Use the parameter *Item* as the mechanism for passing the information by choosing one particular value (or set of values) to represent the fact that a constraint violation has occurred.

(ii) Use an extra parameter to signify that a constraint violation has been detected.

The second approach typically involves an extra parameter called a *state indicator*. Any procedure which can detect a constraint violation is provided with an additional parameter that returns a *status value* which indicates whether or not an error was detected and, if so, what kind of error. The status values are the implementation of the message values in the set M of the specification. In the cursor-based implementation of the **stack** abstract data type the status values might be defined thus:

```
type
    StackStatusType = (Ok, EmptyStack, FullStack);
```

where the status value Ok indicates that no error has been detected. Our implementation of **top** then becomes:

```
procedure Top (S: Stack;  var Item: ItemType;
                          var Status: StackStatusType);
begin
    if IsEmptyStack (S)
    then
        begin
            Writeln (' The stack is empty.');
            Status := EmptyStack
        end
    else
        begin
            Item := S.Item [ S.Top ];
            Status := Ok
        end
end;
```

Exercise 3.13
What changes must be made to the Pascal implementation of **push** to incorporate the use of status values?

Exercise 3.14

When implementing the abstract data type **stack** using a UCSD unit, where should the definition of the status values be placed, and why?

Exercise 3.15

If the rules of Pascal were relaxed to allow any data type to be returned by a function, would you implement the operation **top** as a function?

3.6 Summary of chapter

This chapter has examined the problem of converting a specification into an implementation in Pascal. The conversion is a two stage process: first choose a representation and then write the implementation. The representation determines the data structure and hence the type definitions for the implementation.

There are two principles to be applied when designing an implementation in Standard Pascal.

(i)   Write one procedure for each operation.

(ii) 'Hide' the data structure by defining a type whose name is the same as that of the abstract data type being implemented. For some implementations this may mean parcelling up several structures into a single record.

The objective is to end up with a set of procedures and/or functions, that can be made available to an application programmer, but which have the property that it is impossible to tell, from the headings alone, what representation or implementation has been used.

The discussion then turned to the concept of information hiding which, in short, means: (a) providing the application programmer with only that information about the implementation of the abstract data type which is needed to write an application, nothing more or nothing less, and (b) preventing the application programmer from making use of details of the implementation. Information hiding has the objective of separating the details of an implementation of an abstract data type from an application. This ensures that an implementation can be changed without affecting the application. That is, it ensures that an application is independent of the method of implementation.

The **stack** abstract data type was implemented in UCSD Pascal to see how the principles of information hiding could be applied in a real situation. It was discovered that the use of UCSD **units** was helpful in the quest for complete information hiding but did not provide all necessary facilities. This exercise enabled us to draw up a list of ideal requirements in a programming language to support information hiding. In particular we need to:

(i)   separate the whole of the implementation of an abstract data type from an application program;

(ii) limit access to the implementation of an abstract data type to the headings of the routines which implement the abstract data type operations;

(iii) allow the application program to manipulate more than one instance of the abstract data type without having to replicate the implementation;

(iv) separately compile the implementation of an abstract data type and make the implementation easily available to more than one application program;

(v)   be able to change the implementation of an abstract data type without in any way affecting an application program;

(vi) be able to define the type of the abstract data type item in the application program.

The final section looked at constraint violations, commonly referred to as errors. The discussion considered the impact of constraint violations, recognized by a specification, on an implementation. We concluded that such constraint violations should be dealt with by the application program and that it was the responsibility of the implementation to enable that task to be carried out successfully. Two strategies were proposed.

(i)   Demand that the application should: (a) detect that a particular use of an operation would lead to a constraint violation, and (b) take evasive action to ensure that the event does not occur. The only action required by the implementation is to detect when the application program fails to perform the requisite test and print out a message to this effect.

(ii)   Allow the implementation to detect a constraint violation and report it to the application program so that remedial action can be taken. This strategy requires that additional information be passed between implementation and application, either via a new parameter (a status value) or by using one of the existing parameters.

4 Encapsulation in high level languages

4.1 Introduction

Encapsulation is the term used to describe those features of a programming language which support information hiding. You have seen that standard Pascal offers no support, and that UCSD Pascal falls short of our ideals. There have been many attempts at designing languages which provide encapsulation. In this book we shall examine two such languages: Ada and MODULA-2. Our choice of languages has, to some extent, been a pragmatic one. Both Ada and MODULA-2 have their roots in Pascal and, therefore, should be easy for a Pascal programmer to understand; in fact, you will probably be quite surprised by the similarities among the three languages. It is also our belief that these languages will be around for some time to come. That is not to say that there do not exist other languages eminently suited to our purposes. There is an evolving group of languages, known as *object-oriented languages,* which offer new approaches to program development. We shall look at the concepts behind such languages in a later chapter, but for the present, if you understand the principles behind encapsulation in Ada and MODULA-2, you will be in a good position to appreciate the facilities offered by this new generation of languages. Ultimately, it is not the syntax of a language which is significant; it is more important to have a good understanding of the problems that the language is trying to solve.

Precisely which languages will be seen to be useful in the long term is difficult to predict. Our intention here is to concentrate on the requirements for information hiding as they apply to the implementation of abstract data types. You should not view what follows as an attempt to teach programming in either Ada or

MODULA-2. We shall use your understanding of Pascal as a basis for investigating the general nature of these two languages. In so doing, we shall of course explain those details of the languages necessary for a full understanding of their encapsulation features. As a result you will end up with a sufficient feel for both languages that will enable you to pick up the missing details with relative ease.

There are several points of similarity between Ada and MODULA-2 which distinguish them from Pascal. Both view a program as a collection of similarly constructed parts — in Ada they are called *subprograms,* in MODULA-2 they are *modules* — and collections are said to form a *library.* An application programmer's task is to construct new libraries using, as far as possible, the facilities already available in existing libraries. The advantages of this approach are clear. Re-using already written and tested software can reduce the time taken to produce a new software product with obvious savings in time and cost. It also means that a customer can have more confidence in the correctness of a piece of software. The ability to create useful groups of routines, for a graphics package for example, and to store them in a library, extends the utility of a programming language by making more *tools* immediately available to the applications programmer.

Extending a language through libraries is a well established technique, and languages often include *standard libraries* as part of their design. One commonly occurring standard library is that for input/output routines. I/O has long been a thorny problem for language designers for two reasons. First, the ever increasing diversity of peripheral devices requires different kinds of program statement to control them, and second, varying degrees of software control over I/O are required by programmers. Some languages, such as Pascal, have I/O statements built into them, but it has been recognized that this approach is too inflexible. An alternative is to provide a library of commonly required I/O facilities — a basic set of routines — out of which the programmer can select the most appropriate for a particular application. This scheme offers even greater flexibility when it also permits easy supplementation of the library with new facilities as they become necessary. An additional benefit of the latter approach comes at run-time because only those I/O routines selected by the programmer need be included in the final program. Thus a flexible library system can save on both storage and time requirements of a program. Ada and MODULA-2 both provide I/O facilities via libraries.

Quite clearly, the provision of libraries should be supported by facilities for information hiding because an application programmer does not need to know anything about the implementation of such libraries. Both Ada and MODULA-2 encourage programmers to view their task as the construction of a library in which they:

(i)   obtain, or *import,* certain *resources* from other libraries, and

(ii) identify those objects, or *resources,* in the new library which might be of use to other programmers. In this case they are said to identify the resources which may be *exported* from their library.

In many ways, this view of a library is a generalization of the concept of a procedure with parameters. In a library, however, the resources which can be imported or exported can be types, procedures, or functions as well as values.

Libraries are the main feature by which Ada and MODULA-2 support information hiding. Therefore, we shall examine in detail the *package* facility in Ada and the *module* facility in MODULA-2. To keep you on familiar ground we shall examine the implementation of the **stack** abstract data type in both languages.

4.2 Information hiding in Ada

4.2.1   *Constructing Ada packages*

Ada is one of the many modern languages that has been designed with information hiding in mind, and its major device for encapsulation is called the **package**. A package enables the programmer to group together a set of related routines in a way that hides their implementation. A package is very similar in construction to a UCSD unit in that it has two parts named **specification** and **body**, which correspond to the **interface** and **implementation** parts of a unit, respectively. The significant difference between a package and a unit is that the specification is kept physically separate from the body. You can view the package structure diagrammatically, as shown below.

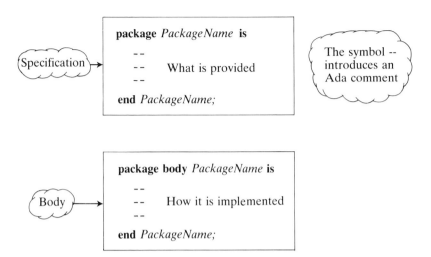

Before a package can be *used* both its specification and its body must exist. However, the two sections can be developed and compiled separately (although the specification must be compiled first).

Several points about Ada syntax and conventions for writing Ada can be identified in this example. Every Ada programming structure has a distinctive keyword to start it and has a corresponding end-of-structure keyword. Thus, every package specification starts with the keyword **package** and terminates with the keyword **end**. The package body starts with **package body** and stops with **end**. Placing the name of the package after the obligatory **end** keyword is optional in each case but is well worth doing. Every Ada construct is terminated with a semicolon, which is why there is a semicolon at the end of both the specification and the body. We shall adopt the typographical convention that Ada reserved words (keywords) will be printed in bold lower case and identifiers will be written in italics with an initial capital letter.

In use, the Ada specification is made available to the applications programmer and contains, amongst other things, the headings of the routines which implement the operations of an abstract data type. The body, which contains the implementation of the routines, is hidden from the applications programmer. As an example, here is an Ada package which implements the **stack** abstract data type. This implementation resembles the UCSD Pascal implementation given in the previous chapter. We shall first display the specification part with the warning that, although it is correct Ada, it suffers from the same deficiencies as the original UCSD unit. We shall, however, seek to rectify the deficiencies later.

```
package EmployeeStack is
    MaxSize: constant Integer := 100;
    subtype Cursor is Integer range 0 .. MaxSize;
    subtype ItemType is String (1 .. 5);
    type Store is array(1 .. MaxSize) of ItemType;

    type Stack is
        record
            Top: Cursor;
            Item: Store;
        end record;

    procedure CreateStack (S: out Stack);
    function IsEmptyStack (S: in Stack) return Boolean;
    function Top (S: in Stack ) return ItemType;
    procedure Pop (S: in out Stack);
    procedure Push ( Item: in ItemType; S: in out Stack);
end EmployeeStack;
```

If you look at the detail of the package specification you will see the close resemblance between Ada syntax and Pascal syntax. There are some differences, of course, which we shall now highlight.

(i)   Ada has a declare-before-use rule which means that declarations of variables, constants, and types must appear before they are used. However, there is no restriction on the order in which definitions of constants, types, and variables must appear provided that the declare-before-use rule is obeyed.

Constants, such as *MaxSize,* are declared with the reserved word **constant** *following* the identifier. The syntax for the declaration of variables (there are none in this specification) is similar to that for constants but without the reserved word **constant**. Had the reserved word **constant** been omitted in the above example then *MaxSize* would have been declared as an integer variable whose initial value was 100.

Note that *every* type declaration must start with either of the reserved words **type** or **subtype**. The use of **type** means that a *totally* new type is being introduced, and variables of that type will be treated as different from all other types of variable. A **subtype** means that variables of this type are compatible with variables of the base type from which the subtype is derived. Hence, had we tried to define *Cursor* as a new **type**, assignments such as *S.Top := 6* would be invalid since *S.Top* is of type *Cursor* whereas *6* is of type *Integer*, and the strong typing in Ada would not permit the assignment.

(ii) Structured types, such as arrays and records, are declared in an almost identical fashion to Pascal. Record definitions start with the keyword **record** and finish with the reserved word **end record**. Array definitions use parentheses to enclose the index range (Pascal uses square brackets).

(iii) There are three **modes** of parameters: **in**, **out**, and **in out**. Roughly speaking, **var** parameters in Pascal correspond to Ada's **in out** mode, and Pascal's value parameters correspond to Ada's **in** mode. The mode of each parameter is specified in the formal parameter list, after the colon but before the type of the parameter. The effects of these modes on the use of the formal parameter in the body of a routine are given in *Table 4.1*.

Table 4.1 Parameter modes in Ada

Effect of mode	**mode**		
	in	**out**	**in out**
Acts as	A local constant.	A local variable.	A local variable.
Value	Provided by the corresponding actual parameter in the procedure call.	Obtained as the result of execution of the procedure, and then assigned to the corresponding actual parameter.	Permits access and assignment to the corresponding actual parameter.
Operations allowed	Input only. Cannot be modified by the routine.	Output only. Can be used and modified by the routine.	Input and output. Can be used and modified by the routine.
Actual parameter	Any expression of the appropriate type.	A variable of the appropriate type.	A variable of the appropriate type.

Notice how the Ada parameter modes reflect the requirements of our specifications. We can, if we so wish, implement source data and results of operations by **in** and **out** parameters, respectively, the latter feature not being available in Pascal. Of course, we can still use **in out** parameters for efficiency purposes.

(iv) To avoid side-effects, a function is allowed parameters of mode **in** only. The type of the result of a function is specified after the reserved word **return** in the function heading. The method of returning the function's value will be discussed later. For now, we shall record that there is no restriction on the type of value that can be returned from an Ada function.

The body of the package *EmployeeStack* has the same basic structure as the implementation part of the corresponding UCSD unit.

```
package body EmployeeStack is
   procedure CreateStack ( S: out Stack);
   begin
      S.Top := 0;
   end CreateStack;

   function IsEmptyStack ( S: in Stack) return Boolean;
   begin
      return ( S.Top = 0);
   end IsEmptyStack;

   function Top ( S: in Stack) return ItemType;
   begin
      if IsEmptyStack (S)
      then
         Put ("The stack is empty.");
      else
         return S.Item (S.Top);
      end if;
   end Top;

   procedure Pop ( S: in out Stack);
   begin
      if IsEmptyStack (S)
      then
         Put ("The stack is empty.");
      else
         S.Top := S.Top - 1;
      end if;
   end Pop;
```

```
   procedure Push ( Item: in ItemType; S: in out Stack);
   begin
     if S.Top = MaxSize
     then
         Put (" The stack is full.");
     else
         S.Top := S.Top + 1;
         S.Item ( S.Top) := Item;
     end if;
   end Push;
 end EmployeeStack;
```

We can now discuss the additional features of Ada which have been introduced in this example.

(v)   The terminating keyword for the if statement in Ada is **end if.** The inclusion of an end of construct keyword has meant that the then– and else-parts are clearly delineated; there is therefore no need to use **begin ... end** constructs, as in Pascal, for compound statements within the if statement.   Note, once again, the use of a semicolon as a *terminator*; each statement is terminated with a semicolon, regardless of what follows in the program. This of course is in direct contrast with Pascal, which uses the semicolon as a separator.

(vi)   The value to be returned by a function is specified in a **return** statement.   There can be several such statements in the body of a function. This avoids having to assign to the function name as in Pascal.

(vii) Unlike UCSD units, the full procedure heading, including formal parameters, has to be specified in the package body.

(viii) The procedures *Get* and *Put* perform interactive text input and output respectively.   They are similar to the Pascal *read* and *write* procedures. However, *Get* and *Put* have one parameter, only.   To read in or write out more than one value means repeated use of *Get* or *Put*.   To output onto a new line use can be made of the *New_Line* procedure.   This procedure has one input parameter to specify the number of lines to be advanced.   Thus, *New_Line(3)* will advance three lines, leaving two blank lines between lines of output.   As we said earlier, modern languages provide input–output facilities via libraries, and this is true of Ada.   The procedures *Get, Put,* and *New_Line* are made available to the programmer from a package named *Text_IO*.   It is incumbent on the programmer, therefore, to declare the wish to make use of this particular package. Precisely how this is done will be discussed later in the section on using packages.

When the body of a package is compiled, the Ada compiler will make use of the type definitions held in the package specification. The specification must, therefore, have been compiled prior to the compilation of the corresponding body. In general, all the components of an Ada program must be compiled into a program library before execution can take place. The rule for compilation is simple: the parts of a program upon which other pieces depend must be compiled first.

Exercise 4.1

Complete the following package specification for a linked representation of a stack in Ada. Pointers in Ada are known as **access** types. Ada has a strict declare-before-use rule which makes the declaration of pointers difficult. Ada overcomes this problem by allowing the programmer to introduce an identifier on its own and to leave the information about its type until later on in the declarations.

```
package EmployeeStack2 is
    subtype ItemType is String(1..5);
    type StackRecord;                    -- declare before use
    type Link is access StackRecord;     -- pointer to stack record
    type StackRecord is                  -- type details come here

                                         fill in the missing details
    type Stack is                        including function and
                                         procedure headings

end EmployeeStack2;
```

The package specifications given so far have suffered from the same problems as the equivalent UCSD units — the implementation of the abstract data type is *usable* in the specification parts of the packages. This problem has been recognized in the design of Ada through the provision of **private** types. In *Exercise 4.1* you were introduced to the fact that Ada allows you to separate out the introduction of an identifier from its type definition. We can use this device to hide the details of the implementation. To see what must be done, consider a copy of the type definitions within the specification of *EmployeeStack*.

```
package EmployeeStack is
    MaxSize: constant Integer := 100;
    subtype Cursor is Integer range 0 .. MaxSize;
    subtype ItemType is String (1 .. 5);
    type Store is array (1 .. MaxSize) of ItemType;
    type Stack is record
                    Top: Cursor;
                    Item: Store;
                end record;
```

From all this information the user only really needs to know that the *ItemType* is *String;* the remaining information concerns only the implementation and can be declared to be private, as we see next, in the revised package specification.

```
package EmployeeStack is
    subtype ItemType is String(1 .. 5);
    type Stack is private;

    procedure CreateStack (S: out Stack);
    function IsEmptyStack (S: in Stack) return Boolean;
    function Top (S: in Stack) return ItemType;
    procedure Pop (S: in out Stack);
    procedure Push (Item: in ItemType; S: in out Stack);

private
    MaxSize: constant Integer := 100;
    subtype Cursor is Integer range 0 .. MaxSize;
    type Store is array (1 .. MaxSize) of ItemType;
    type Stack is record
                    Top: Cursor;
                    Item: Store;
                end record;
end EmployeeStack;
```

There are several important points to note about this revised specification.

(i) The identifiers *ItemType* and *Stack* have to be declared (i.e. introduced) before they are used in the formal parameters of the routines that follow them.

(ii) The type of the identifier *Stack* has been declared to be private which means that its type will be defined in the private part of the specification.

You may well be tempted to say that this revision has not solved the problem; after all the implementation details are still there in the package specification and are, therefore, visible to the applications programmer. This is true. However, the private declarations are effectively local to the package specification: the user of the package will not be allowed to make use of these definitions in any program. Thus, the implementation is visible but **not** accessible. That part of the specification *not* declared as private gives the information which is available externally to the package. Hence, outside the package, variables and constants of type *Stack* can be declared and the five procedures/functions used as required.

Exercise 4.2

Revise the package specification for *EmployeeStack2* to take advantage of the private type to improve its information hiding properties.

The accessible parts of the package specifications *EmployeeStack* and *EmployeeStack2* are identical, thus confirming that we have achieved our information-hiding objective of divorcing *what* is required from *how* it is implemented. If an implementation is to be altered, the changes may be required in both the package body and the private types of the package specification. That part of the package specification accessible to the user will not be changed. Hence the user's program will not need to be changed, although it will need recompiling since part of the package specification has been altered.

Quite clearly, the Ada solution goes a long way to meeting our ideals of information hiding. However, the use of private types in the specification has its drawbacks. Ideally we should prefer to have the contents of the private part lodged within the package body. The designers of Ada had to compromise between efficiency and complete invisibility of an implementation which, in the current state of the art, is difficult for a compiler to achieve.

Check point

(i)   What are the components of an Ada package?

(ii) In general terms, when implementing an abstract data type using an Ada package, what is placed in the package specification and what is placed in the package body?

(iii) What is the private part of a package specification used for?

(iv) In what order must the components of a package and an application program be compiled?

(v) How is I/O provided in Ada?

(vi) What are the parameter modes in Ada?

(vii) In an Ada package employing private types in its specification what is likely to need changing if the method of implementation of the abstract data type is altered?

Solution

(i)   A package has two components: a specification part and a body.

(ii) The package specification contains:
        the type definition of the items;
        the headings of the procedures/functions which implement the abstract data type operations;
        the type definitions of the data structure (declared to be private).
      The package body contains:
        the implementation of the procedures/functions which implement the abstract data type operations.

(iii) The private declarations part of a package specification holds the definition of the data structure used in the implementation of the abstract data type.

(iv) The package specification is compiled first, followed by the package body. Finally the applications program can be compiled.

(v) I/O is provided via libraries. Interactive I/O is provided by the *Text_IO* package.

(vi) Ada has three parameter modes: **in**, **out**, and **in out**. The first is used for input only parameters, the second for output only parameters, and the third is for parameters used for both input and output.

(vii) Only the private declarations will need modification. The package specification will need re-compiling.

4.2.2   Using Ada packages

Gaining access to the resources in an Ada package is straightforward. As an example, suppose that an applications programmer wishes to make use of our package *EmployeeStack* in writing a subprogram (i.e. a procedure or function). Placing the statement

> **with** *EmployeeStack*;

in front of the subprogram heading causes the Ada system to *import* (make available) the *EmployeeStack* package into the new subprogram. Thereafter, in this subprogram, reference to the visible resources (types, procedures, functions, and so on), can be made using 'dot-notation'. For example, the following two statements

> *EmployeeStack.Push* ("*Susan* ", *NameStack*);

> *CurrentName* := *EmployeeStack.Pop* (*NameStack*);

respectively push and pop an item onto and off a stack named *NameStack*. You will notice that the procedure names *Push* and *Pop* are prefixed with their package name. This device ensures that, in circumstances where two or more packages are imported into a subprogram, there can be no confusion between resources with the same name in different packages. In cases where the applications programmer is satisfied that there can be no such ambiguity the **use** statement can make references to imported resources less cumbersome. If, immediately preceding the applications subprogram, we wrote:

> **with** *EmployeeStack*; **use** *EmployeeStack*;

we could then, for example, within the subprogram, write:

    Push ("Susan ", NameStack);

and

    CurrentName := Pop (NameStack);

(the **use** statement in Ada is similar to Pascal's with-statement, but is in force for the whole of the subprogram that it precedes).

Here is an example of the use of the *EmployeeStack* package in a subprogram to write out a list of names in the reverse order of their input. As the subprogram uses I/O it also has to use the *Text_IO* package.

```
with EmployeeStack; use EmployeeStack;
with Text_IO; use Text_IO;
procedure Revs is
   NameStack: Stack;                    -- an instance of a stack is declared
   CurrentName: ItemType;
   EndOfListMarker: constant ItemType := "ZZZZZ";
   Count: Integer := 0;                 -- initializing a variable
begin
   CreateStack (NameStack);             -- an instance of a stack is created
   Put ("Enter first name:");
   Get (CurrentName);
   while not (CurrentName = EndOfListMarker)
   loop
      Push (CurrentName, NameStack);
      Count := Count + 1;
      Put ("Enter next Name:");
      Get (CurrentName);
   end loop;
   for J in 1 .. Count  loop            -- the loop parameter, J, is
      Put (Top (NameStack));            -- implicitly declared here, and
      Pop (NameStack);                  -- cannot be accessed outside loop
      New_Line;
   end loop;
end Revs;
```

There are some further points about Ada which have been illustrated in the above subprogram.

(i)   Variables can be initialized within their declaration.

(ii) The body of a loop starts with the reserved word **loop** and ends with the reserved word **end loop**. Unless preceded by a **while** or **for**

construct, a **loop ... end loop** construct will cause a set of statements to
be repeated indefinitely.

(iii) A **for** loop implicitly declares its loop control variable (called a
loop parameter in Ada). Such a parameter must not be declared in the
declarations part of the subprogram, nor must it be referenced outside
the loop. Also, the loop parameter may not be modified (i.e. assigned
to) within the loop.

4.2.3   *Generic packages*

The subprogram *Revs* made use of a single stack. That instance of a
stack was created using the *EmployeeStack* operation *CreateStack*.
For this operation to work, a variable of type *Stack* had to be declared
in the application *Revs,* using the type definition *Stack* imported from
the package *EmployeeStack*. Quite clearly, an application subprogram
can be written which seeks to manipulate several such stacks. This
would be achieved by declaring a variable for each stack and then
applying *CreateStack* to each variable, before performing any of the
other stack operations.

Provided that the application program is to manipulate stacks in
which the *ItemType* is always *String,* all is well. However, it is
conceivable, and it happens in practice, that an application program will
need to manipulate several stacks and that those stacks will have
different types of item in them. One way to do this, which is
unacceptably inefficient, is to create several copies of the stack package
each with its own (different) *ItemType*.

Clearly, this could lead to maintenance problems since any
modification of the package would have to be carried out on all copies.
Also, it will not be readily apparent to a reader that the packages are
almost identical: the details of the package body would need to be
compared to determine this. The Ada *generic* facility provides a
solution to this problem by first allowing a *template* for a package,
called a **generic package**, to be defined. A generic package can then be
used to create occurrences of the package, known as generic
instantiations, as we explain below.

Here is a generic package for stacks that we have named
StackPackage; it differs very little from our original *EmployeeStack:*

```
generic
    type ItemType is private;          -- ItemType is a generic parameter

package StackPackage is          -- we've changed the name - it's a new package
    type Stack is private ;
    procedure CreateStack ( S: out Stack);
    function IsEmptyStack ( S: in Stack) return Boolean;
    function Top ( S: in Stack) return ItemType;
    procedure Pop ( S: in out Stack);
    procedure Push ( Item: in ItemType; S : in out Stack);

private
    MaxSize: constant Integer := 100;
    subtype Cursor is Integer range 0 .. MaxSize;
    type Store is array(1 .. MaxSize ) of ItemType;
    type Stack is   record
                        Top: Cursor;
                        Item: Store;
                    end record;
end StackPackage;
```

The type definition for *ItemType,* which now comes immediately after the reserved word **generic**, is a *generic parameter.* Such generic parameters are a mechanism for specifying, outside the package in some other subprogram, the types of certain objects used within the package. For example, suppose that in an application program a stack called *NumberStack* is required, containing a stack of integers. Then the application program will contain the following declaration:

```
package NumberStack is new StackPackage (Integer);
```

The effect is to bring into existence a new package called *NumberStack. NumberStack* is a specific instance of *StackPackage* in which the type of *ItemType* has been set to *Integer.* Thereafter, in the application program, reference can be made to the subprograms *NumberStack.CreateStack, NumberStack.Push, NumberStack.Pop,* and so on. Of course, the use of

```
use NumberStack;
```

in the application will avoid having to prefix the stack operations with the package name, *NumberStack.* The package *NumberStack* which results from this instantiation behaves as a normal directly written out package. In particular, it exports the type *Stack* which can be used in an applications program to declare several instances of stacks (each with items of the type defined in the **new** construct).

The **generic** facility enables general packages to be written for which certain objects have their types defined at a later stage. As another example of this, suppose we want to construct a generic stack package in which, not only do we want to leave the type of the stack items to be specified by the applications programmer, we would also like the programmer to specify the maximum number of items that can be stored in the stack. The specification part of a suitable package is given below.

```
generic
    type ItemType is private;           -- ItemType is a generic parameter
    MaxSize : Natural;                  -- so is MaxSize

    package AnotherStackPackage is      -- we've changed the name again
        type Stack is private;

    procedure CreateStack ( S: out Stack);
    function IsEmptyStack ( S: in Stack) return Boolean;
    function Top ( S: in Stack ) return ItemType;
    procedure Pop ( S: in out Stack);
    procedure Push ( Item: in ItemType; S: in out Stack);

private
    subtype Cursor is Integer range 0 .. MaxSize;
    type Store is array (1 .. MaxSize ) of ItemType;
    type Stack is record
                    Top: Cursor;
                    Item: Store;
                 end record;
    end AnotherStackPackage;
```

The object *MaxSize* has been converted to a generic parameter of type *Natural*. Objects of type *Natural* can take integer values which are positive or zero (but not negative). The value which will eventually be associated with *MaxSize* will be provided when the package is instantiated. For example, the declaration

package *NumberStack* **is new** *AnotherStackPackage (Integer, 50);*

will create a stack package in which the items must be Integer and whose stacks can contain at most 50 items.

Exercise 4.3

What changes are needed to the package *EmployeeStack2* (defined in *Exercise 4.1*) to turn it into a generic stack package?

Throughout this discussion we have concentrated on the changes that have to be made to the specification part of a package to ensure that the package conforms, as far as Ada will allow, to the concept of information hiding. No changes have had to be made to the corresponding package bodies. Thus, the package body associated with the generic package *StackPackage* has the same contents as that of the package body for *EmployeeStack* given in *Section 4.2.1*. Clearly the name of the package body must correspond to that of the package specification.

The Ada facility for defining a generic package means that it is possible to take a generic specification of an abstract data type and make it available to various applications. The (generic) parameters of the generic package enable it to be tailored to a particular application. The use of private types means that access to the data structure is restricted to the specified operations implemented as procedures and/or functions in the package body.

Check point

For each of the six ideal requirements for encapsulation given at the end of *Section 3.4*, say whether Ada meets each requirement and, if so, what features of the language support that requirement.

Solution

(i)   The implementation of an abstract data type is separated completely from the application. The implementation is held in the **body** of a package.

(ii) Access to the abstract data type is limited to the headings of the routines which implement its operations. The headings are contained in a package **specification**. Certain types and routines which are part of the implementation may have to appear in the specification but, by declaring them **private**, the application program cannot have access to them. Such objects will, however, be visible to the application programmer and this may influence the design of the application program.

(iii) It is possible to declare more than one instance of an abstract data type. By defining the abstract data type as a type within the **specification** part of a package, the type is available for use in the application program, and several variables of that type can be declared.

(iv) The **specification** and the **body** of a package are compiled separately. They are also compiled independently of the application program. The **with** statement is used to make a package available to an application program.

(v) Changes to an implementation come in two forms:
    (a) changes to the implementation of the bodies of the procedures/functions which reside in the **body** of a package. Such changes can be made without affecting the applications program in any way.
    (b) changes to the **private** part of a package **specification**. The application program will have to be recompiled but no changes in its source code will have been necessary.

(vi) **Generic packages** enable the type of an abstract data type to be defined in the application program.

4.3 Information hiding in MODULA-2

MODULA-2 is a direct descendant of Pascal. It was designed and developed by the same person who invented Pascal: Niklaus Wirth. Indeed, MODULA-2 emerged, after careful design deliberations, as a language that includes all aspects of Pascal and is extended by the inclusion of the important concept of the module. In this account we shall assume a knowledge of Pascal, and we shall concentrate on the new module concept. We shall also assume that you have studied the previous subsection on information hiding in Ada. The module concept has much in common with packages in Ada, and it will be useful to compare the two approaches.

The structure of a module looks, at first glance, very much like a Pascal main program.

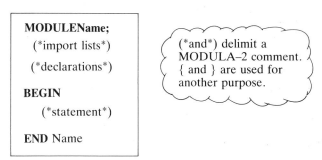

Here is a simple example showing how closely MODULA-2 resembles Pascal. The example illustrates the use of standard libraries and the role of import lists. It calculates the area and circumference of a number of circles; the radius of each circle is input from the terminal.

Note : At the time of writing, the MODULA-2 language was not officially standardized. The most recently published report on the language is contained in the book *Programming in MODULA-2* by Niklaus Wirth. As a result there are a variety of implementations of the language which are incompatible. The modules given in this book conform to the requirements of the *MacMeth* compiler (version 2.3, 1987), a system written for the Apple Macintosh computer by members of the Institut für Informatik, ETH Zurich (including Niklaus Wirth). All versions of MODULA-2 insist that keywords are written in upper case.

```
MODULE Circle;
    FROM InOut IMPORT ReadCard, WriteString, WriteLn;
    FROM RealInOut IMPORT ReadReal, WriteReal;

    CONST Pi = 3.14159265;
    VAR Count, Limit: CARDINAL;    (* non-negative integers *)
        Circumference, Area, Radius: REAL;

BEGIN
    Count := 0;
    WriteString ('Enter number of circles:');
    ReadCard (Limit);
```

```
REPEAT
    Count := Count + 1;
    WriteString ('Enter radius'); ReadReal (Radius);
    Circumference := 2 * Pi * Radius;
    Area := Pi * Radius * Radius;
    WriteString ('Circumference = ');
    WriteReal (Circumference, 15);
    WriteLn;
    WriteString ('Radius = ');
    WriteReal (Radius, 15);
    WriteLn
UNTIL Count >= Limit
END Circle.
```

There are several points to note about this example.

(i)   This is a complete program that can be compiled and executed.

(ii) As in Ada, MODULA-2 has a set of standard libraries which
contain routines for I/O, mathematical functions, and so on. The library
module *InOut* contains a set of useful routines for input and output (I/O
for reals is separated out into the library module *RealInOut*). The
mathematical library module *MathLib0* contains such basic
mathematical routines as *sqrt, exp, ln, sin,* and *cos*.

(iii) To make use of objects defined in other modules, a module must
import them. That is, at the start of a module there must be a list of all
external objects required by the module. The **FROM** statement is the
mechanism by which objects are imported. So, for example, the
statement

   FROM *InOut* **IMPORT** *ReadCard, WriteString, WriteLn;*

makes available the routines *ReadCard, WriteString,* and *WriteLn*
from the module named *InOut*. Thus, only the routines required by a
module need to be imported. If all the routines in a module are required
then the shorter statement

   IMPORT *InOut;*

is all that is required. As with Ada, there is always the possibility that
there will be similarly named objects in different modules, and the same
'dot notation' is used to resolve the ambiguity. For example, the name
InOut.WriteString refers to the object named *WriteString* contained

in the module *InOut*. Prefixing an object's name with the name of the module in which it is defined is known as a *qualified import* and, when used, the shortened form of the **IMPORT** statement is mandatory.

(iv) In general terms, a module's ability to import an object from another module will only succeed if that other module explicitly *exports* that object. To export an object, that is, to make it available for other modules to import the object, an **EXPORT** statement can be used. Such a statement is placed immediately after any import statements. Definition modules (of which more later) do not require explicit **EXPORT** statements. Exporting will be considered in more detail later.

(v) The type **CARDINAL** is equivalent to Ada's *Natural* type. That is, variables of type **CARDINAL** can take non-negative integer values.

For the purposes of information hiding, a module can be split into two parts called a **definition module** and an **implementation module.** These are the direct equivalents of the package specification and package body in Ada. Definition modules and implementation modules can be compiled separately, provided that a definition module is compiled prior to its corresponding implementation module. Here is a definition module for our **stack** abstract data type.

```
DEFINITION MODULE StackOps;
    TYPE ItemType = ARRAY [ 0..80 ] OF CHAR;
    TYPE Stack ;        (* this is known as opaque export: see below *)

    PROCEDURE CreateStack ( VAR S: Stack);
    PROCEDURE IsEmptyStack ( S: Stack ):  BOOLEAN;
    PROCEDURE Top ( S: Stack): ItemType;
    PROCEDURE Pop ( VAR S: Stack );
    PROCEDURE Push ( Item: ItemType; VAR S: Stack);
END StackOps.
```

You should note several points from this example.

(i) MODULA-2 retains Pascal's parameter mechanism. Parameters are passed by value or by reference, and in the latter case the formal parameter is preceded by the keyword **VAR.**

(ii) Functions are allowed in MODULA-2 but they are introduced by the keyword **PROCEDURE.** In the example, *IsEmptyStack* and *Top* are functions. There are restrictions on the type of value returned by functions (all simple types can be returned).

(iii) There is no standard string type in MODULA-2. Some string I/O operations, *WriteString,* for example, are available which consider a string to be defined as

   ARRAY [*0 .. 80*] **OF CHAR**

Some implementations of MODULA-2 provide string handling modules as standard. (c.f. Pascal and UCSD Pascal).

(iv) The type of the object *Stack* has been hidden. It will be specified in the corresponding implementation module. This technique is called *opaque export.*

(v) All objects defined in a definition module are exportable, hence there is no need for an export list.

(vi) A definition module can have an import list, and can define constants, types, variables, and procedures. All objects so defined are available for use in the corresponding implementation module without having to be explicitly imported into that implementation.

   Here is the corresponding implementation module for *StackOps.* We have chosen to use a linked representation. You will notice that this module imports two objects named *Allocate* and *Deallocate* from the module *System. Allocate* and *Deallocate* are equivalent to the Pascal procedures **new** and **dispose**; *System* is a standard module. *Allocate* requires the use of the standard function, **SIZE**, which returns the amount of storage space needed for the new object.

```
IMPLEMENTATION MODULE StackOps;
   FROM InOut IMPORT WriteString;
   FROM System  IMPORT Allocate, Deallocate;

   TYPE
      Link = POINTER TO StackRecord;
      Stack = Link;
      StackRecord =  RECORD
                        Item: ItemType;
                        Previous: Link;
                     END RECORD;
```

```
PROCEDURE  CreateStack ( VAR S: Stack);
BEGIN
   Allocate (S, SIZE (StackRecord));
   S := NIL
END CreateStack;

PROCEDURE  IsEmptyStack (S: Stack): BOOLEAN;
BEGIN
   RETURN (S = NIL)
END IsEmptyStack;

PROCEDURE  Top ( S: Stack): ItemType;
BEGIN
   IF IsEmptyStack (S)
   THEN
      WriteString ('The stack is empty.');
   ELSE
      RETURN S ↑.Item
   END
END Top;

PROCEDURE  Pop ( VAR S: Stack);
VAR Temp: Link;
BEGIN
   IF IsEmptyStack (S)
   THEN
      WriteString (' The stack is empty.');
   ELSE
      Temp := S ;
      S := S ↑.Previous;
      Deallocate (Temp)
   END
END Pop;

PROCEDURE  Push ( Item: ItemType; VAR S: Stack);
VAR P: Link;
BEGIN
   Allocate (P, SIZE (StackRecord));
   P ↑.Item := Item;
   P ↑.Previous := S;
   S := P;
END Push;
END StackOps.
```

You should recognize the majority of constructs in this example: they are little changed from the Pascal originals. Note however, there are two important points about MODULA-2.

(i) The definition of the object *Stack* is given in full in the implementation module.

(ii) Whenever an object name is introduced in the definition module but the object type is left to be defined in the implementation module (i.e. opaque export), that object is restricted to being of type pointer or to subranges of standard types. In particular, record and array types *cannot* be used in opaque export. This is an awkward restriction since we cannot use opaque export in our cursor-based implementation of the stack in the normal way. However, there is a solution: in an implementation, the type identifier used for the abstract data type is defined as a *pointer to* the required data structure thereby making the name of the abstract data type available for opaque export. In doing this, one has to be careful to remember, in the implementation of the procedures, that one is dealing with a pointer variable and not the usual record or array.

Here is an example of the use of the module *StackOps* in a MODULA-2 main module.

```
MODULE Reverse;
FROM StackOps IMPORT Stack, ItemType, CreateStack, Push, Top, Pop;
FROM InOut IMPORT WriteString, ReadString, WriteLn;
CONST EndOfListMarker = 'ZZZZZ';
VAR NameStack: Stack;          (* an instance of a stack is declared *)
    CurrentName: ItemType;
    Count, J: Cardinal;
BEGIN
    CreateStack (NameStack);    (* an instance of a stack is created *)
    Count := 0;
    WriteString ('Enter first name: ');
    ReadString (CurrentName);
    WHILE NOT (CurrentName = EndOfListMarker) DO
        Push (CurrentName, NameStack);
        Count := Count + 1;
        WriteString ('Enter next Name: ');
        ReadString (CurrentName);
    END;
    FOR J := 1 TO Count DO
        WriteString (Pop (NameStack));
        WriteLn
    END;
END Reverse.
```

Check point

For each of the six ideal requirements for encapsulation given at the end of *Section 3.4,* say whether MODULA-2 meets each requirement and, if so, what features of the language support that requirement.

Solution

(i)   Modules separate the implementation of an abstract data type from application programs.

(ii) Applications programs have access to the contents of the definition module only. All objects defined in a definition module are available for export.

(ii) The applications program can declare several instances of an abstract data type provided the data structure is introduced in the definition module. Ideally the data structure should be defined as an opaque export.

(iv) Modules are compiled separately.   An **IMPORT** statement (in an applications program) makes the objects in a definition module available for use in that applications program.

(v)   Changes to an implementation occur in the **IMPLEMENTATION** module which can be changed without affecting the applications module. The **IMPLEMENTATION** module, if changed, must be recompiled. This does not affect either the **DEFINITION** module or the applications program.

(vi) There are no generic modules in MODULA-2.

4.4 Exception handling

Ada offers an alternative method of dealing with constraint violations known as *exception handling*.   Ada's mechanism allows a package to detect a constraint violation and to communicate this fact to the user without the need for additional parameters. Instead, objects called *exceptions* are used. For example, here is the package specification for our cursor-based *EmployeeStack* abstract data type that includes exceptions.

```
package EmployeeStack is

    type ItemType is String;
    type Stack  is private ;
    StackFull, StackEmpty: exception;              -- exceptions

    procedure CreateStack ( S: out Stack);
    function IsEmptyStack ( S: in Stack) return Boolean;
    function Top ( S: in Stack) return ItemType ;
    procedure Pop ( S: in out Stack);
    procedure Push ( Item: in ItemType; S: in out Stack);

private
    MaxSize: constant Integer := 100;
    type Cursor is range 0 .. MaxSize;
    type Stack  is   record
                Top: Cursor;
                Item: array(1 .. MaxSize) of ItemType;
              end record;
end EmployeeStack;
```

The only change is the addition of the declaration of two variables *StackEmpty* and *StackFull* of type *Exception*. In the same way that the procedures *CreateStack, IsEmptyStack, Top, Pop,* and *Push* are made available for use in another subprogram so are the two exceptions. The exceptions are to be used to indicate when attempts have been made to access an item in an empty stack and to push an item on to an already full stack, respectively. Now the way in which exceptions are used is quite different from the use of normal variables. When using exceptions we are said to be *handling* them. To see how *exception handling* is done in Ada, here is the *Revs* application modified for use with the revised *EmployeeStack*. The new part of the code is at the end of the program.

```
with EmployeeStack;  use  EmployeeStack;
with Text_IO;  use  Text_IO;
procedure Revs  is
    NameStack: Stack;
    CurrentName: ItemType;
    EndOfListMarker: constant  Itemtype := "ZZZZZ";
    Count: Integer := 0;
begin
    CreateStack (NameStack);
    Put ("Enter first name: ");
    Get (CurrentName);
    while not (CurrentName = EndOfListMarker)
    loop
        Push (CurrentName, NameStack);
        Count := Count + 1;
        Put ("Enter next Name: ");
        Get (CurrentName);
    end loop;
    for J  in  1 .. Count  loop
        Put (Pop (NameStack));  New_Line;
    end loop;

exception                       -- exception handling starts here
    when StackFull =>           -- exception handlers are always at the end
        -- do something

    when StackEmpty =>
        -- do something else
end Revs;
```

The original version of *Revs* did not attempt to take special action if a constraint violation occurred. The revised version has an extra exception handling part added. In the new piece of code there are two *exception handlers* — pieces of program, introduced by the keyword **when**, which are executed only if their respective exceptions have occurred. That is, whenever an exception is detected (by *EmployeeStack*) the appropriate exception is *raised* and, no matter where in the applications program (*Revs*) execution is, control passes to the appropriate handler. This means that the programmer has full control over what is to happen when an exception occurs. For example, the programmer might simply choose to print out a message and stop processing. Alternatively, if the exception arises because incorrect data was input, the handler could be programmed to ask the user to re-input the data and then resume processing.

In principle, exception handling in Ada does not offer any new programming capability. As we remarked in *Section 4.2.2,* one strategy for dealing with constraint violations is to guard each call to a package procedure. The exception handling process would then be incorporated into the guard as shown below.

```
if IsEmptyStack (TheStack)
then
    -- do something (the exception handler)
else
    Top (TheStack)
end if;
```

The argument for Ada's mechanism is that such guards obscure the normal flow of control through a program, making the program difficult to read and comprehend. Since constraint violations occur only infrequently (or at least that is what *should* happen), by removing the exception handling to a separate part of the program the remaining part of the program is left to deal with 'normal' processing. There is much debate on this topic, and time will tell which approach gains majority approval.

So far, we have discussed how exceptions are the means of communicating the fact that a constraint violation has occurred, and how exceptions are used in an applications program. It remains to be shown how an exception is *raised* in the first place. Here is the *EmployeeStack* package body showing how exceptions are raised.

```
package body EmployeeStack

    procedure CreateStack ( S: out Stack);
    begin
        S.Top := 0;
    end CreateStack;

    function IsEmptyStack ( S: in Stack) return Boolean;
    begin
        return ( S.Top = 0)
    end IsEmptyStack;

    function Top ( S: in Stack) return ItemType;
    begin
        if IsEmptyStack (S)
        then
            raise StackEmpty;                    -- empty stack exception raised
        else
            return S.Item (S.Top);
        end if;
    end Top;
```

```
procedure Pop (S: in out Stack);
begin
   if IsEmptyStack (S)
   then
      raise StackEmpty;              -- empty stack exception raised
   else
      S.Top := S.Top – 1;
   end if;
end Pop;

procedure Push ( Item: in ItemType; S: in out Stack);
begin
   if S.Top = MaxSize
   then
      raise StackFull;              -- full stack exception raised
   else
      S.Top := S.Top + 1;
      S.Item (S.Top) := Item;
   end if;
end Push;
end EmployeeStack;
```

All that has happened is that the messages 'The stack is empty' and 'The stack is full' have been replaced by **raise** statements. At the start of execution all exceptions are deemed not to have occurred, and it is through the raise mechanism that an exception is set. Once the exception has been handled the exception is unset again.

We have described only the simplest use of exceptions. There are inbuilt exceptions in Ada to cover such situations as divide-by-zero and attempting to use an array subscript which is out of range. The exception handling mechanism gives the programmer the opportunity to deal with these situations instead of being forced to accept the system default (which is usually to abort execution). Another important aspect of exception handlers deals with the situation in which an application procedure uses a package that includes exceptions but fails to provide appropriate exception handlers. Here, the Ada system looks for an appropriate handler (one that mentions the exception's name) in the subprogram which invoked the application procedure. If no handler is found then a search is made of the next enclosing subprogram. Eventually the system will look in the main program and, if a handler is still not found, the default action is taken.

In the case of large programs there can be a large number of exceptions and, as with **case** statements, you might like to deal with several exceptions in the same way, or deal with a few exceptions individually and then deal with all that remain in a uniform way. Ada allows you to do both. For example, you can write

```
exception
    when StackFull | StackEmpty =>
        -- do something
```

and also:

```
exception
    when StackFull =>
        -- do something
    when others =>                 -- handles everything other than the
        -- do something else       -- exceptions explicitly mentioned
                                    -- above. USE WITH CARE.
```

Although our example has not shown it, exceptions can be used with all Ada subprograms, including generic packages.

Exercise 4.4

Describe the alterations that would have to be made to the *EmployeeStack2* package (see *Exercise 4.1*) in order to incorporate exceptions as the method for dealing with constraint violations.

Check point

(i)   How does Ada's exception handling mechanism help with the implementation of constraint violations?

(ii) What, in Ada, is an exception?

(iii) What is an exception handler?

(iv) What is the purpose of a **raise** statement?

(v)   What three things must be provided in an Ada program to make use of exception handling?

(vi) Into which of the two strategies for dealing with constraint violations does Ada's exception handling mechanism fall?

Solution

(i) A constraint violation causes an exception to be raised which is then dealt with, in the application program, by a piece of program called an exception handler.

(ii) An exception is a variable which is set to signal that some unusual event has occurred and is subsequently acted upon by an exception handler.

(iii) An exception handler is a piece of program which is executed once an exception has been raised.

(iv) A **raise** statement is used to set an exception.

(v) To make use of exception handling, the program must include
  (a) the declaration of exception variables;
  (b) **raise** statements to set the exceptions;
  (c) exception handlers to be executed when exceptions have been raised.

(vi) Exception handling deals with constraint violations *after* the events have occurred.

4.5 Summary of chapter

This chapter has examined the encapsulation facilities in both Ada and MODULA-2. The Ada package provides almost all the necessary facilities apart from making the abstract data type's data structure completely invisible. Ada supports generic packages.

MODULA-2, through its modules, supports total hiding of the implementation of an abstract data type. Its two drawbacks are the awkward limitation on the types that can take part in opaque export, and the lack of generic modules.

Packages and modules have the same general structure (as with UCSD units). A package, module or unit has two parts: one containing the definition of objects that can be used by an applications program, the other containing the implementation of these objects. Packages (and modules) physically separate the specification (definition) part from the body (implementation) part for the purpose of separate compilation.

Ada functions are not restricted in the type of value they can return. MODULA-2 functions are restricted in the type of value that can be returned, but the restriction is not as severe as in Pascal.

Ada's **exception handling** facility enables information about the occurrence of constraint violations to be passed between packages and applications programs without the use of parameters. The mechanism permits the applications programmer to write pieces of code to deal with exceptions when they arise. The exception handlers are separated from the normal processing by being placed at the end of a subprogram. The fact that an exception has been detected is signalled by *raising* the exception.

5 The abstract data type queue

5.1 Introduction

In this chapter we consolidate the preceding material by examining the specification, representation, and implementation of the abstract data type **queue**. As with a stack, we assume that you are already familiar with a queue. In a fashion similar to our account of a stack in *Chapter 1,* here is a brief and informal description of a queue.

> *A queue is a collection of data kept in sequence. Each item of data is of the same type. Data are added only at one specific end of the sequence (usually called the tail); whilst data are deleted and retrieved only at the other end of the sequence (usually called the head).*

This description can be illustrated with a pictorial example of a queue.

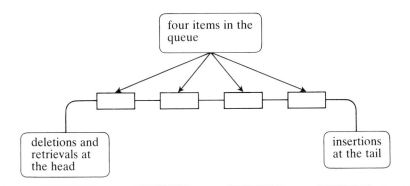

Fig. 5.1 A queue with four items

117

We shall assume that the abstract data type queue is defined by the following five operations:

createqueue *returns a new empty queue;*

front *takes a queue and returns the item at the head of the queue as its result when the queue is not empty, or the message 'the queue is empty' when the queue is empty;*

addtoqueue, *given an item and a queue, returns a queue with that item inserted at the tail;*

deletefromqueue *takes a queue and returns, as its result, that queue with the item at the head deleted when the queue is not empty, or the message 'the queue is empty' when the queue is empty;*

isemptyqueue *returns true when the queue is empty, or false when the queue is not empty.*

5.2 The specification of the abstract data type queue

5.2.1 *A constructive specification*

Fig. 5.2 gives constructive specification of a queue, using the underlying model of a list, developed in *Chapter 2*.

NAME
  queue (item)

SETS
  Q   set of queues
  I   set of items
  B   set consisting of the Boolean values *true* and *false*
  M   set of message values consisting of the single member *the queue is empty*

SYNTAX

    createqueue: $\rightarrow$ Q
    front: Q $\rightarrow$ I $\cup$ M
    addtoqueue: I $\times$ Q $\rightarrow$ Q
    deletefromqueue: Q $\rightarrow$ Q $\cup$ M
    isemptyqueue: Q $\rightarrow$ B

SEMANTICS

    pre-**createqueue**( ) ::= *true*
    post-**createqueue**(q) ::= q = **createlist**
    pre-**front**(q) ::= *true*
    post-**front**(q; r) ::= *if* q = **createlist**
                     then
                          r = *the queue is empty*
                 else
                          r = **first**(q)
    pre-**addtoqueue**(i,q) ::= *true*
    post-**addtoqueue**(i, q; r) ::= (**concatenate**(q, **make**(i)))
    pre-**deletefromqueue**(q) ::= *true*
    post-**deletefromqueue**(q, r) ::= *if* q = **createlist**
                         then
                            r = *the queue is empty*
                     else
                            r = **trailer**(q)
    pre-**isemptyqueue**(q) ::= *true*
    post-**isemptyqueue**(q: b) ::= b = **isemptylist**(q)

invariant assertion

    Whenever an operation is applied to a value from M then the result of the operation is that same value from M.

Fig. 5.2 The constructive specification of a queue

There are alternative expressions that can be substituted in this specification. For example, an alternative for *q = **createlist*** is *isemptylist(q)*.

The ease of production and readability of this constructive specification demonstrates the efficacy of the constructive approach once a suitable underlying model has been selected (and defined).

To conclude this section on the constructive specification, we now consider how the *generic* specification of *Fig. 5.1* may be used to produce a *particular* specification with slightly different semantics.

Exercise 5.1

There is a need to specify the abstract data type *xqueue*, which is a specific application of the generic abstract data type *queue*, with slightly changed semantics. The abstract data type *xqueue* includes all five of the operations defined above. For convenience of distinction, they are named **createxqueue**, **frontx**, **addtoxqueue**, **deletefromxqueue** and **isemptyxqueue**. The semantics have been changed slightly in two respects. *First,* a size constraint has been placed on the number of items in the *xqueue*. Should an attempt be made to add more than 100 items to the *xqueue*, then the message *the xqueue is too full* results. *Second,* a new operation, **sizewarning**, has been introduced. When applied to a non-empty *xqueue*, this operation returns the message *the xqueue is not getting full,* when there are less than 75 items in the *xqueue,* or the message *the xqueue is getting full,* when there are 75 items or more in the *xqueue*. The action of **sizewarning** on an empty *xqueue* is undefined. Write down the constructive specification of the abstract data type *xqueue*.

5.2.2 An axiomatic specification

The axiomatic specification of a queue is now given, partly by way of contrast, but also as an introduction to more complex issues (than those raised by a stack) about axiomatic specifications in general. *Fig. 5.3* gives the axiomatic specification: the *NAME, SETS,* and *SYNTAX* entries have been omitted, since they are the same as in *Fig. 5.2*.

SEMANTICS
$\forall\, i \in I,\ \forall\, q \in Q:$

    isemptyqueue(createqueue) = *true*   *(Q1)*

    isemptyqueue(addtoqueue(i, q)**)** = *false*   *(Q2)*

front(**createqueue**) = *the queue is empty* (*Q3*)
front(**addtoqueue**(i,q)) = *if* **isemptyqueue**(q)
$\qquad$ *then*
$\qquad$ i
$\qquad$ *else*
$\qquad\qquad$ **front**(q) (*Q4*)

deletefromqueue(createqueue) = *the queue is empty* (*Q5*)
deletefromqueue(addtoqueue(i,q)) =
$\quad$ *if* **isemptyqueue**(q)
$\quad$ *then*
$\qquad$ **createqueue**
$\quad$ *else*
$\qquad$ **addtoqueue**(i,deletefromqueue(q)) (*Q6*)

invariant assertion
$\quad$ *Whenever an operation is applied to a value from M then the result of the operation is that same value from M.*

Fig. 5.3 The axiomatic specification of a queue

The issues about axiomatic specifications that we will now examine centre around two features: *recursive definitions* and the manner in which the axioms *relate the meanings* of operations to one another. Several of the axioms are unproblematic in terms of the explanation given for the axiomatic specification of a stack in *Chapter 2*. For example, *axioms Q1* and *Q2* can be seen as straightforwardly defining the outcome of **isemptyqueue** when applied to an empty queue and to a queue that is not empty, respectively. However, the other axioms are not quite so straightforward in two respects. First, *axiom Q4* and *axiom Q6* are recursive definitions. Second, it can be seen that the axioms do not *explicitly* define the meaning of an operation. For example, the meaning of **addtoqueue** is defined implicitly by relating its meaning to **front** (*Q4*), and to **deletefromqueue** (*Q6*). In particular, it is not clear from a cursory glance at the axioms that **addtoqueue** and **deletefromqueue** operate at different ends of the queue. These issues will be explained by considering the manner in which the axioms define the outcome of some specific applications of the operations.

First, we consider the application of **deletefromqueue** to the queue depicted in *Fig. 5.4*.

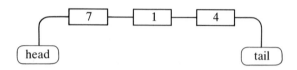

Fig. 5.4 The queue before **deletefromqueue**

Intuitively, such an application of **deletefromqueue** should produce the queue depicted in *Fig. 5.5*.

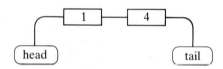

Fig. 5.5 The queue after **deletefromqueue**

How may this intuition be shown formally from the axioms of *Fig. 5.3?* Only two axioms, *Q5* and *Q6*, involve the use of **deletefromqueue**. Of these, *axiom Q5* is not immediately relevant, since it addresses the issue of **deletefromqueue** applied to an empty queue. Therefore, the formalization must use *axiom Q6*. The key to using *Q6* lies in writing the application, to the queue in *Fig. 5.4*, of **deletefromqueue** in the form of the left-hand side of *Q6*. Let *y* denote a queue containing two items. The item at the head has a value of *7*, whilst the item at the tail has a value of *1*. Rather than represent this queue in a pictorial fashion, we write it using the list notation introduced in *Chapter 2*. That is, *y* denotes the queue

<7, 1>

with the head at the left and the tail at the right.

addtoqueue(4, y)

is an expression that evaluates to the queue used as source data for the application of **deletefromqueue** — the queue depicted in *Fig. 5.4.* This application of **deletefromqueue** may be written as

deletefromqueue(addtoqueue(4, y))   *(1)*

which is in the form of the left-hand side of *axiom Q6*. The axiom states that *(1)* is the same as

if **isemptyqueue(y)**
then
   createqueue
else
   addtoqueue(4, deletefromqueue(y))   *(2)*

That is, the required result is obtained by

addtoqueue(4, deletefromqueue(y))   *(3)*

The nature of the recursive definition can be seen since, from *(3)*,

deletefromqueue(y)   *(4)*

needs to be evaluated. Using a similar strategy *(4)* may be written in the form of the left-hand side of *axiom Q6* as

deletefromqueue(addtoqueue(1, x))   *(5)*

where x denotes the queue

<7>

Using the right-hand side of *axiom Q6 (5)* may be written as

if **isemptyqueue(x)**
then
   createqueue
else
   addtoqueue(1, deletefromqueue(x))   *(6)*

Hence,

deletefromqueue(x)   *(7)*

needs to be evaluated, remembering that its result must then be used to go back and evaluate *(6)*, which in turn will then be used to go back and evaluate *(3)*. Using the same strategy as before *(7)* may be written in the form of the left-hand side of *axiom Q6* as

deletefromqueue(addtoqueue(7, createqueue))   *(8)*

since **addtoqueue(7, createqueue)** results in the queue x. Using the right-hand side of *Q6* expression *(8)* is the same as

if **isemptyqueue(createqueue)**
then
   createqueue
else
   addtoqueue(7, deletefromqueue(createqueue))   *(9)*

Thus *(7)* and *(8)* evaluate to the result of **createqueue**, that is, an empty queue. So *(6)* becomes (the *else* is the only relevant part)

addtoqueue(1, createqueue)   *(10)*

and, using *(10)*, expression *(3)* becomes (again, the *else* is the only relevant part)

addtoqueue(4, (addtoqueue(1, createqueue)))   *(11)*

which is the queue <1, 4>, thus formalizing the intuition of *Fig. 5.5*.

As well as demonstrating the use of a recursive definition more complex than **length**, the above example illustrates how the meanings of operations are related to one another. *Axiom Q6* does not *say* explicitly that **addtoqueue** adds items at the tail of a queue and that **deletefromqueue** removes them from the tail. However, it *does* relate the two operations to one another, so that it is clear that **addtoqueue** does the opposite to **deletefromqueue** and, furthermore, does it at the *opposite* end of the queue. Ignoring the condition of an empty queue, *axiom Q6* can (very) roughly be paraphrased as *deleting something from a queue to which an item has just been added is the same as deleting from the original queue and then adding the item*. This rendition, of course, ignores the recursive nature of the axiom. You should notice that the queue on the right-hand side — denoted by *deletefromqueue(q)* — is smaller than the queue on the left-hand side — denoted by *addtoqueue(i, q)* . Repeated applications of the axiom eventually result in the left-hand side evaluating to the case covered by the condition of an empty queue, hence terminating the recursion.

Exercise 5.2

Show how the axioms of *Fig. 5.3* define the result of **front** applied to the queue depicted in *Fig. 5.4* to be the item whose value is 7.

Check point

(i) *q = createlist* and *isemptylist(q)* are two expressions that are alternatives for each other. What type of result does each expression yield?

(ii) Which axiom, in *Fig. 5.3,* defines the action of **front** on a non-empty queue?

Solutions

(i) Both expressions evaluate to a Boolean.

(ii) *Axiom Q4.*

5.3 A linked representation

A queue can be represented as a linked list of items with links pointing to the ends of the list. The inclusion of an extra record, known as the *dummy* record, simplifies the implementations of inserting and deleting

queue items, as you will see. Here is a picture of such a queue
representation showing three items (all character values).

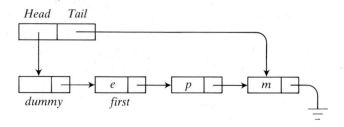

Fig. 5.6 Representation of a queue

An empty queue (as would exist after the application of the operation
createqueue) can be represented by the dummy record alone as shown in
Fig. 5.7.

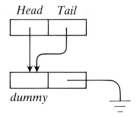

Fig. 5.7 An empty queue

An empty queue can therefore be recognized by the fact that both the
head and *tail* links point to the same record (the dummy). You should
be quite clear that the *head* and *tail* links are pointing to the **whole** of
the dummy record and not, as it may appear at first sight, to the
individual fields of that record. The fact that we keep the **nil** link in the
tail record will subsequently help with deletions.

5.4 A pointer-based implementation in UCSD Pascal

A linked representation can be implemented as a UCSD Pascal unit whose interface part is:

```
unit QueueOps;
interface
   type
      ItemType = Char;       {or whatever the application demands}
      Link = ↑QueueRecord;
      QueueRecord = record
                        Item: ItemType;
                        Next: Link;
                     end;
      Queue = record
                  Head, Tail: Link
               end;
   procedure  CreateQueue (var Q: Queue);
   function IsEmptyQueue (Q: Queue ): Boolean;
   procedure  AddToQueue (Item: ItemType; var Q: Queue);
   procedure  Front (Q: Queue; var Item: ItemType);
   procedure  DeleteFromQueue ( var Q: Queue);
```

The most significant point about this implementation is that a queue is accessed via two pointers: *Head* and *Tail*. The type *Queue* must therefore be implemented as a record. This implies that, due to the restrictions on the values that can be returned from a Pascal function, the majority of queue operations have to be implemented as procedures. We have also taken certain other decisions.

(i)   Each queue operation has been implemented by one procedure or function.

(ii) The operations **addtoqueue** and **deletefromqueue** have a queue as both source data and result. We have chosen to implement them as a single **var** parameter.

(iii) The type of the items held in the queue is character. This can be changed quite easily by amending the definition of *ItemType*: no other changes are necessary.

(iv) Constraint violations are dealt with by printing out the appropriate message. Since the function *IsEmptyQueue* is provided, an applications programmer will have the necessary tools for detecting possible constraint violations.

The implementation of the queue operations proceeds with *CreateQueue*, and the situation depicted in *Fig. 5.7* can be achieved with

```
procedure CreateQueue (var Q: Queue);
var
   Temp: Link;
begin
   New (Temp);                     {create a new queue record }
   Temp.Next := nil;               { put nil into new queue record }
   Q.Head := Temp ;                { make Head point to new record }
   Q.Tail := Temp                  { make Tail point to new record }
end {CreateQueue };
```

Our implementation of **isemptyqueue** is

```
procedure IsEmptyQueue (Q: Queue): Boolean;
begin
   IsEmptyQueue := (Q.Head = Q.Tail )
                            {both Head and Tail point to same record }
end {IsEmptyQueue};
```

To see how to add an item to a queue, *Fig. 5.8* illustrates the state of a queue both before and after an item is added.

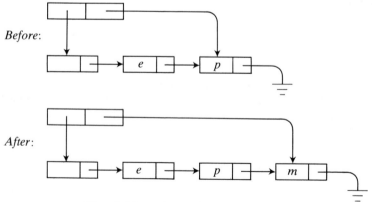

Fig. 5.8 Inserting an item into a queue

Here the new item, *m*, is placed in a new record added to the end of the list. Thus, the required actions are:

(i) create a new record, and add the new data item (*m*) to it;

(ii) link the new record to the tail of the queue by amending (a) the *Tail* pointer, and (b) the pointer in the *Next* field of the previous tail record (containing *p*).

In Pascal this process can be written as

```
procedure AddToQueue (Item: ItemType; var Q: Queue);
var
   Temp: Link;
begin
   New (Temp);                  {Create a new record}
   Temp.Item := Item;           {Add item to new record}
   Temp.Next := nil;            {Place nil into link field of new record}
   Q.Tail↑.Next := Temp;        {New record is added to end of list}
   Q.Tail := Temp               {Point to new tail record}
end {AddToQueue};
```

Exercise 5.3

What amendments, if any, have to be made to *AddToQueue* to add an item to an empty queue?

The next operation to be implemented is **front**. Here, we only need to access the item held in the first record (pointed to from the dummy) provided, of course, that the queue is not empty.

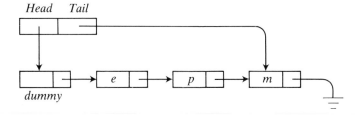

Fig. 5.9 Accessing the first item in a queue

The procedure is

```
procedure Front (Q: Queue; var Item: ItemType);
begin
  if IsEmptyQueue (Q)
  then
    writeln ('The queue is empty: cannot get front item.')
  else
    Item := Q. Head↑.Next↑.Item
end {Front};
```

For the implementation of **deletefromqueue**, we again look at the situation before and after the operation is carried out on a typical queue

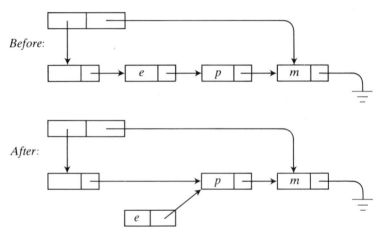

Fig. 5.10 Deleting an item from a queue

Deleting an item from a queue means removing the item at the head of the queue. The only action required is to arrange that the pointer field of the dummy item points to the same item as the pointer field of the first record. Thus the first item (*e* in the example) will be by-passed and, in normal circumstances, the dummy will end up pointing at what was the second record. There is one awkward situation which arises when there is only one item in the queue. When this happens, the first item is also the last item, and the *Tail* pointer must be modified accordingly, as shown in *Fig. 5.11*.

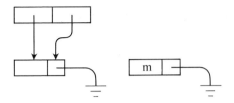

Fig. 5.11 Deleting the only item (*m*), in a queue

Here is the procedure

```
procedure DeleteFromQueue ( var Q: Queue);
var
    Temp: Link;
begin
  if IsEmptyQueue (Q)
  then
    writeln ('The queue is empty: cannot delete.')
  else
    begin
      Temp := Q.Head ↑.Next;          {points to first record}
      Q.Head ↑.Next := Temp ↑.Next;   {update dummy pointer field}
      if Temp ↑. Next = nil
      then                            {deleting last record}
        Q.Tail := Q.Head;
        dispose  (Temp)
    end
end {DeleteFromQueue};
```

Exercise 5.4 draws your attention to the advantages of using a dummy record in the implementation.

Exercise 5.4

In broad terms, what changes would have to be made to the implementation of the queue described above if a dummy record were not used?

5.5 A cursor-based representation

In this section we shall present a cursor-based representation which is not immediately implementable in most programming languages, particularly Pascal. This example will illustrate that the most natural representation for an abstract data type may require much additional work in order to implement it. We shall examine the *circular structure* in *Fig. 5.12*.

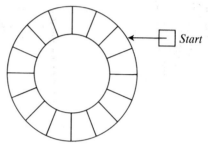

Fig. 5.12 A circular structure

This structure has a known, fixed number of storage elements, each capable of holding a single character. One of the elements is designated the *start* element, and there is one operation **movecursor** which takes as its source data a cursor (an indicator to a storage element) and returns as its result a cursor at the next storage element in order around the structure; it does not matter which way **movecursor** goes — clockwise or anticlockwise — so long as the movement is consistent in one direction. Suppose *current* is a cursor, then *Fig. 5.13* shows the effect of applying the operation **movecursor** to *current*.

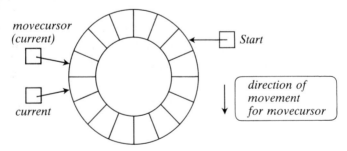

Fig. 5.13 The effect of **movecursor**

The next diagram, *Fig. 5.14*, shows a queue with four elements (after several additions and deletions have taken place). The *Tail* pointer will always point to the next available slot.

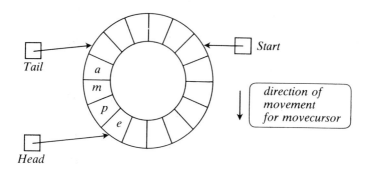

Fig. 5.14 A queue stored in a circular structure

This representation is useful because the queue operations are easily achieved. The first item in the queue is accessed through the *Head* cursor; a new item is added by storing it at the next position after the current *Tail* position, and deleting an item means moving the *Head* cursor by one position. A circular structure is a finite (static) structure which means we have to take steps to check when the queue occupies the whole of the structure to avoid overwriting existing entries when adding new items. As with the linked representation, it is often a useful strategy to incorporate a dummy item to help with the awkward cases: empty queues and queues with just one element in them. An empty queue is then represented by setting both the *Head* and *Tail* cursors to point to the dummy item.

Exercise 5.5

If the circular structure incorporates a dummy item and the *Head* cursor is made always to refer to the dummy item, how can the conditions that the structure is (i) empty, and (ii) full, be detected?

5.6 Implementation in UCSD Pascal

The first step in implementing the abstract data type queue is to *assume*
the existence of the circular structure (and its operation *MoveCursor*) in
our programming language. That is, we have the definitions and
declarations in *Fig. 5.15* available to us.

```
type Queue = record
                 Head, Tail: Cursor;
                 CS: CircularStructure
             end;

function MoveCursor (C: Cursor): Cursor;
```

Fig. 5.15 Definitions for circular structure implementation

Precisely how we shall implement a cursor and the storage for the
elements of the circular structure will be postponed until we have
implemented the queue in terms of the circular structure. We begin with
createqueue.

```
procedure  CreateQueue (var Q: Queue);
begin
   Q.Head := Start;
   Q.Tail := Start
end {CreateQueue};
```

The next two functions determine whether the queue is empty or full.

```
function  IsEmptyQueue (Q: Queue): Boolean;
begin
   IsEmptyQueue := (Q.Head = Q.Tail)
end {IsEmptyQueue};

function  IsFullQueue (Q: Queue): Boolean;
var
   Next: Cursor;
begin
   Next := MoveCursor (Q.Tail);
   IsFullQueue := (Next = Q.Head)
end {IsFullQueue};
```

The following procedure obtains the item at the front of the queue. Precisely how items stored in a circular structure are accessed will depend on the nature of our implementation of the structure. Since we have not specified this yet we shall assume the existence of a procedure *GetItem* which takes as input both a circular structure and a cursor value (indicating one of the structure's elements), and returns the value of the item stored in that element.

```
procedure Front (Q: Queue; var Item: ItemType);
begin
    GetItem (Q.CS, Q.Head, Item)
end {Front};
```

The operation **addtoqueue** is implemented as

```
procedure AddToQueue (Item: ItemType; var Q: Queue);
begin
  if IsFullQueue (Q)
  then
      writeln ('The queue is full: cannot add.')
  else
      begin
        PutItem (Q.CS, Q.Head, Item);              {see below}
        Q.Tail := MoveCursor (Q.Tail)
      end
end {AddToQueue};
```

Again we have had to invent a procedure, *PutItem*, that will add an item to a specified location in the circular structure. The final procedure implements **deletefromqueue**.

```
procedure DeleteFromQueue ( var Q: Queue);
begin
  if IsEmptyQueue (Q)
  then
      writeln ('The queue is empty: cannot delete')
  else
      Q.Head := MoveCursor (Q.Head)
end {DeleteFromQueue};
```

We have successfully implemented the queue using the circular structure without worrying too much about the way in which the circular structure will be provided. In fact, we only had to invent two new procedures, *GetItem* and *PutItem*, to deal with accessing items in the circular

structure. It remains, therefore, to choose some appropriate storage
mechanism for the circular structure and to implement the three missing
routines, *PutItem, GetItem,* and *MoveCursor.*

The most obvious way to store the circular structure is in an array and
to use index values as cursors to indicate where in the array the head and
tail records are located. *Fig. 5.16* shows the situation following the
insertion of three (character) items into a queue (including a dummy
record at the tail).

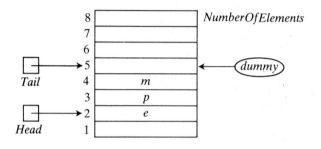

Fig. 5.16 Circular structure implemented as an array

With this implementation we can write down the following definitions
(which should be read in conjunction with the definitions in *Fig. 5.15*).

```
const
    NumberOfElements = 100;                    {or whatever is appropriate}
    Start = 1;
type
    Cursor = 1 .. NumberOfElements;
    CircularStructure = array[ Cursor ] of ItemType;
```

We can now implement the remaining procedures and functions,
beginning with *MoveCursor.* In general, *MoveCursor* must increment
an index value by 1, unless the index is *NumberOfElements,* in which
case the index must be set to 1. There are several ways to achieve this,
and we present just one.

```
function MoveCursor (C: Cursor): Cursor;
begin
  MoveCursor := 1 + (C mod NumberOfElements)
end {MoveCursor};
```

Here are the two access procedures.

```
procedure  GetItem (CS: CircularStructure; C: Cursor;
                        var Item: ItemType);
begin
  Item := CS [ C]
end {GetItem};
```

```
procedure  PutItem ( var  CS: CircularStructure; C: Cursor;
                        Item: ItemType);
begin
  CS [ C] := Item
end {PutItem};
```

The whole of this implementation can be packaged up into a UCSD unit.

```
unit QueueOps;
interface
  const
    NumberOfElements = 100;               {or whatever is appropriate}
    Start = 1;
  type
    ItemType = Char;                      {or whatever is appropriate}
    Cursor = 1 .. NumberOfElements;
    CircularStructure = array[ Cursor ] of ItemType;
    Queue =   record
                  Head, Tail: Cursor;
                  CS: CircularStructure
              end;

  procedureCreateQueue (var Q: Queue);
  procedure IsEmptyQueue  (Q: Queue): Boolean;
  procedure IsFullQueue (Q: Queue): Boolean;
  procedure Front (Q: Queue;  var Item: ItemType);
  procedure AddToQueue (Item: ItemType; var Q: Queue);
  procedure DeleteFromQueue ( var Q: Queue);
```

implementation

```
function MoveCursor (C: Cursor): Cursor;
begin
   MoveCursor := 1 + (C mod NumberOfElements)
end {MoveCursor};


procedure  GetItem (CS: CircularStructure; C: Cursor;
                          var Item: ItemType);
begin
   Item := CS [ C]
end {GetItem};


procedure  PutItem ( var CS: CircularStructure; C: Cursor;
                          Item: ItemType);
begin
   CS [ C] := Item
end {PutItem};


procedure CreateQueue;
begin
   Q.Head := Start;
   Q.Tail := Start
end {CreateQueue};


function IsEmptyQueue;
begin
   IsEmptyQueue := (Q.Head = Q.Tail)
end {IsEmptyQueue};


function IsFullQueue;
var
   Next: Cursor;
begin
   Next := MoveCursor (Q.Tail);
   IsFullQueue := (Next = Q.Head)
end {IsFullQueue};


procedure Front;
begin
   Get (Q.CS, Q.Head, Item)
end {Front};
```

```
procedure AddToQueue;
begin
  if IsFullQueue (Q)
  then
    writeln ('The queue is full: cannot add.')
  else
    begin
      PutItem (Q.CS, Q.Head, Item);
      Q.Tail := MoveCursor (Q.Tail)
    end
end {AddToQueue};

procedure DeleteFromQueue;
begin
  if IsEmptyQueue (Q)
  then
    writeln ('The queue is empty : cannot delete.')
  else
    Q.Head := MoveCursor (Q.Head)
  end {DeleteFromQueue};
end {QueueOps}.
```

The most significant point to notice here is that the routines *MoveCursor, PutItem,* and *GetItem* have been placed in the implementation part of the unit because they are required only by the implementation of the queue operations. In this way, we have hidden these routines from the applications programmer.

Exercise 5.6
Briefly describe how you would incorporate, into *QueueOps*, a routine for use by applications programmers which prints out the current contents of a queue.

Exercise 5.7
Write, in Pascal, the routine referred to in *Exercise 5.6.*

Check point

(i) Why are the two links, *Head* and *Tail,* used in the representation of a queue?

(ii) How is an empty queue represented?

(iii) In the unit *QueueOps*, how can the type of item held in a queue be altered?

(iv) How are constraint violations handled in *QueueOps?*

(v) How can an applications program avoid a constraint violation by *QueueOps?*

(vi) What operations define the *circular structure?*

(vii) Why are the operations for the circular structure placed in the implementation part of the *QueueOps* unit?

Solutions
(i)   Operations on queues take place at both ends of a sequence. Having the two links allows quick access to either end.

(ii) An empty queue is represented by the *Head* and *Tail* links pointing to the same record (the dummy).

(iii) The type of item can be altered by changing the definition of *ItemType*.

(iv) Constraint violations are handled by printing out a message from within the appropriate procedures.

(v) Constraint violation can be avoided by use of the function *IsEmptyQueue,* prior to invoking either of the procedures *Front* or *DeleteFromQueue,* and by use of *IsFullQueue,* prior to invoking *AddToQueue.*

(vi) The *circular structure* is defined by **movecursor, getitem,** and **putitem**. Whilst not explicitly stated, there must also be some mechanism to initialize a circular structure (such as **createcircularstructure**) which serves to indicate the **start** item.

(vii) The circular structure operations are part of the implementation and are not to be made available to an applications programmer.

5.7 Summary of chapter

The abstract data type queue was represented as a linked sequence together with cursors that pointed to both ends of the sequence.

You have seen two different implementations of a queue in UCSD Pascal. The first was a direct pointer implementation, which included a dummy item to facilitate easy manipulation of an empty queue as well as a queue with only one item.

The second implementation involved a *circular structure*. Here, the queue operations were implemented in terms of the operations defining a circular structure. The circular structure was itself implemented using an array. This example illustrated that it may sometimes be convenient, or easy, to represent an abstract data type in terms of data structures not supported by the chosen programming language. In this case, the latter structure will have to be implemented in terms of data structures that *are* supported.

6 The abstract data type deque

6.1 Introduction

The format of this chapter is different from the preceding chapters because it asks you to perform the specification, representation and implementation of an abstract data type *for yourself.* We shall provide an informal description of the abstract data type *deque* (pronounced 'deck') and provide you with hints and suggestions about the approach you should adopt. To gain full benefit from this chapter you should work through the exercises in the order in which they appear since they are designed to help you through all the stages from specification to implementation.

Our specification of the deque is based on the specification of the abstract data type queue given in the previous chapter. If you find that our informal description of a deque is deficient, you will have to decide for yourself what action to take, just as you would do when defining an abstract data type for yourself.

6.2 The specification of the abstract data type deque

The term **deque** comes from the phrase *d*ouble *e*nded *que*ue which gives a useful, if succinct, description of the object. That is, a deque behaves like a queue in which insertions and deletions can occur at *either* end. Thus, a deque is a sequence for which the operations of insertion, deletion, and retrieval are defined, but these operations are constrained to take place only at the ends of the sequence. The ends of a deque are conventionally referred to as the *left* and the *right*. A deque can, therefore, be pictured as shown in *Figure 6.1* in which the boxes represent individual items in the deque.

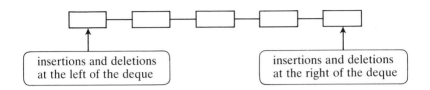

Fig 6.1 A deque, showing the left and right items

Here are the informal descriptions of the eight operations which define the abstract data type **deque**.

createdeque: returns a new, empty deque as its result.

isemptydeque: takes a deque as source data and returns the value *true* when the deque is empty, or the value *false* otherwise.

produceleft: takes a deque as source data and returns the item at the left of the deque when the deque is not empty, or the error value *error — empty deque,* when the deque is empty.

produceright: as for **produceleft,** except that the retrieval takes place at the right of the deque.

addleft: takes an item and a deque as source data and returns, as its result, a deque with the item inserted at the left.

addright: as for **addleft,** except that the insertion takes place at the right of the deque.

deleteleft: takes a deque as source data, deletes the item at the left and returns the changed deque when the source deque is not empty, or the error value *error — empty deque* when the source deque is empty.

deleteright: as for **deleteleft,** except that the deletion takes place at the right of the deque.

Exercise 6.1
If you regard the left of a deque as corresponding to the head of a queue, and the right of a deque as corresponding to the tail of a queue, how do the operations on a deque relate to the operations on a queue?

Exercise 6.2
Write down the *NAME* and the *SETS* components of a formal specification of the abstract data type **deque**.

Exercise 6.3
Write down the *SYNTAX* for a formal specification of the abstract data type **deque**.

Exercise 6.4
Use the constructive approach to the specification of the *SEMANTICS* of the abstract data type **deque**, using the underlying model of a list. HINT: look at the list operations in *Section 2.5.2*.

6.3 A linked representation

Exercise 6.5
Draw a diagram showing a linked representation of a deque with three items. Indicate on the diagram the left and right items of the deque. What is your representation for an empty deque? What does your representation look like with only one item in it?

Exercise 6.6
Show the three element deque devised for *Exercise 6.5* after one application of the **addleft** operation. Which links were updated as a result of this operation?

Exercise 6.7
Show the three element deque devised for *Exercise 6.5* after one application of the **deleteleft** operation. Which links were updated as the result of this operation?

Exercise 6.8

Repeat *Exercise 6.7,* but this time apply the **deleteright** operation.

6.4 A pointer-based implementation

Exercise 6.9

Write down the Pascal type definitions required to implement your representation of a **deque**.

Exercise 6.10

Write down a Pascal procedure that implements the **createdeque** operation.

Exercise 6.11

Write down a Pascal function which implements the **isemptydeque** operation. Check that your function works correctly for a deque with precisely one item in it (i.e. in addition to the dummy record, if one is being used).

Exercise 6.12

Implement the operation **produceleft** as a Pascal procedure.

Exercise 6.13

Write down a Pascal procedure which implements the **addleft** operation. HINT: look at the solution to *Exercise 6.6.* Do not forget that this operation must cater for the addition of an item both to an empty deque and to a deque with items already present.

Exercise 6.14

Write down a Pascal procedure which implements the **deleteleft** operation.

The operations **produceright** and **addright** are implemented in similar fashion to **produceleft** and **addleft** respectively, and so will not be considered further. The implementation of the operation **deleteright** is, however, not as straightforward as any of the other operations. The reason is easy to see. If you look back at the solution to *Exercise 6.8,* you will see that two links have to be updated whenever there are two or more items in the deque: the *Right* pointer in the deque record and, more significantly, the *Link* field of the item which points to the right item. In order to update the latter pointer it is necessary to refer to the penultimate item. The only link to the penultimate item is given by the item to its left. In other words there is no simple access to the penultimate item in the deque.

There are several ways to overcome this difficulty. The most obvious way is to keep another pointer, in addition to the left and right pointers. The extra pointer would point to the penultimate item in the deque. This method requires a change to the implementation of the deque, which we shall avoid. Instead, we shall devise a more involved procedure for the implementation of **deleteright**.

If you begin with the left item, it is possible to follow the *Next* pointers of successive items until the right item is reached. Visiting each item in turn is known as *traversing* the deque. The usual method of traversing a sequence is to use a pointer variable *CurrentItem,* say, which points to the item currently being visited. The statement:

CurrentItem := CurrentItem↑.Next

moves this pointer on to the next item in the sequence. If we also have a pointer variable, *PreviousItem,* say, which always points to the item preceding the current item, then when the current item is the right item, *PreviousItem* will refer to the penultimate item, as required for the deletion operation.

Exercise 6.15
Use the method outlined above to implement the operation **deleteright**. HINT: when there is only one item in the deque there is no need to traverse the deque.

The following exercises ask you to implement the abstract data type deque on your own computer.

Exercise 6.16
Write down either the interface part of a UCSD Pascal unit, or a definition module in MODULA-2, for your implementation of the abstract data type **deque**. Use integer items.

Exercise 6.17
(i) Devise either the implementation part of unit, or the implementation module, corresponding to the interface part/definition module given in the solution to *Exercise 6.16*.

(ii) Enter into your own computer, and debug, your implementation of the abstract data type **deque**.

(No solution is provided for this exercise.)

Exercise 6.18
Devise an applications program which uses the abstract data type **deque**. For example, write a program to do the following:

(i) enter, from the keyboard, 4 items (use integers) into a deque; enter each item on the left;

(ii) retrieve (i.e. obtain copies of) the items, one at a time, from the right and add each one on the right of a new deque;

(iii) retrieve and print out the items in the second deque, from the left.

CHECK: the items should be printed in the order in which they were entered.

Exercise 6.19
Write a procedure to print out the contents of a deque traversing from left to right.

Exercise 6.20
How would you amend the representation of a deque to facilitate the easy printing of deque items by performing the traversal from right to left?

Exercise 6.21

(i) Change the representation of a deque to a cursor based representation, and then re-implement the abstract data type deque.
(No solution is provided for this part of the exercise.)

(ii) Do the applications programs, written for the solutions to *Exercises 6.18* and *6.19,* require modification before they will execute successfully with this revised implementation?

7 Binary search trees

7.1 Trees, binary trees, and binary search trees

Trees crop up in a variety of guises in everyday life, usually for representing relationships between things. Well known examples are: management structures showing responsibility relationships within firms, book classification schemes for libraries, and knock-out tournaments. Indeed, any hierarchical structure can be represented in the form of a tree.

Trees are our first example of an abstract data type that is not based on the notion of a sequence. Its specification (and implementation) will involve the use of recursion.

A tree is a data structure which has zero, one, or more *nodes* organized in a hierarchical manner. Each node contains some data (of the same type) and is joined to other nodes in the certain ways.

(i) Unless the tree is empty there is one special node, the *root,* at the top of the hierarchy.

(ii) Every node, apart from the root, is joined by a *branch* to *exactly* one node at the next higher level in the hierarchy. (A node may be joined to more than one node at the next lower level).

(iii) There is a known, fixed relationship between data items on adjacent levels.

A node which is not joined to another node at a lower level is called a *leaf* node. *Fig. 7.1* shows an example of a tree which has been annotated to illustrate the new terminology.

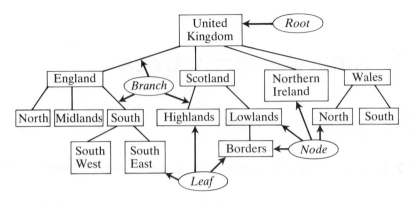

Fig. 7.1 Terminology of trees

For any given node, the node joined to it at the higher level is its *parent,* while the node(s) joined to it at the lower level are its *children.*

Exercise 7.1
Answer the following questions about the tree in *Fig. 7.1.*

(i)   How many leaf nodes are there?
(ii) How many children does the root node have?
(iii) Which node is the parent of the node containing the item *Midlands?*
(iv) What are the child nodes of the node containing the item *Midlands?*
(v)   What are the children of the node containing *Lowlands?*
(vi) What is the parent of the node containing *United Kingdom?*

In a general tree a node can have an unrestricted number of children. In the example above, the largest number of children of a single node is 4, belonging to the node *United Kingdom.* If we restrict to two the maximum number of children that any one node may have then we have a **binary tree**. *Fig. 7.2* shows an example of a binary tree in which there are nodes with zero, one, and two children, but no node has more than two children.

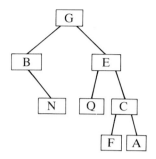

Fig. 7.2 A binary tree

If you were to remove the root node from the tree given in *Fig. 7.1* by cutting the branches leading from the root node, you would end up with the situation shown in *Fig. 7.3*.

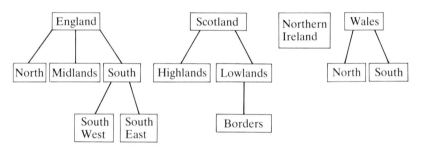

Fig. 7.3 Subtrees of the tree in Fig. 7.1

That is, you would end up with four *subtrees* — one for each child of the root node. The significant point to note here is that a subtree is itself a tree, and has all the properties of a tree.

Exercise 7.2

For the tree on the left hand side of *Fig. 7.3* identify:

(i)   the root node;
(ii) the leaf nodes;
(iii) the children of the root node;
(iv) the number of subtrees that can be obtained from this tree.

Exercise 7.3

Which of the trees in *Fig. 7.3* are binary trees?

Binary trees are the most useful trees because it is possible to transform other kinds of trees into them. There is an important characteristic of a binary tree which makes it very useful for sorting and searching. Each node, apart from the leaf nodes, is the root node of a subtree, and there are precisely two subtrees hanging from that node called the *left subtree* and the *right subtree*. We can arrange to store data in the nodes of a tree in such a way that *all* the data in a left subtree come before the data in the root node, and *all* the data in the right subtree come after the data in the root node. *Fig. 7.4* illustrates such a tree.

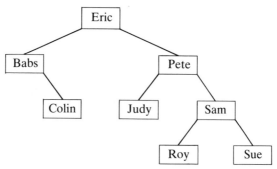

Fig. 7.4   A binary tree with sorted data

You should verify for yourself that, no matter which node you choose, all the data in its left subtree come alphabetically before the name in that node, and that all data in the right subtree comes alphabetically after that name. For example, the root node of the whole tree contains the name *Eric*. The left subtree contains the names *Babs* and *Colin* whereas the right subtree contains the names *Judy, Pete, Roy, Sam,* and *Sue*. The tree in *Fig. 7.4* is an example of a **binary search tree** (BSTree, for short).

To see how such a tree can be constructed, here is an unordered set of names

Eric, Pete, Judy, Babs, Sam, Sue, Roy, Colin,

Starting with an empty tree, add the first name (*Eric*) as the root node.

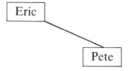

The second name to be added is *Pete*. If you compare the new name with the name in the root node you will discover that *Pete* comes alphabetically after *Eric*. Therefore, we attempt to add the new name to the *right* subtree. Since the right subtree is empty the new name is added as the right child of *Eric*.

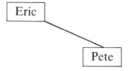

The next name to be added is *Judy*. We again start the process at the root node. Since *Judy* comes after *Eric* it must be added to the right subtree. Therefore, we move from the root node (*Eric*) by following the right branch to the right child (*Pete*). Now the new name (*Judy*) is compared with the current node (*Pete*). As *Judy* comes before *Pete* we add the new node to the *left* subtree of *Pete*. Since the left subtree is empty *Judy* can be added as a new leaf node.

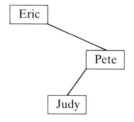

The next name is *Babs* which, coming before *Eric,* is added to the left subtree of *Eric*.

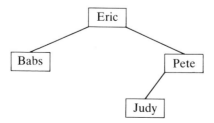

Next *Sam* is added. *Sam* comes after *Eric,* so the right branch from
Eric is taken. *Sam* comes after *Pete,* so attempt to take the right
branch again. As there is no right branch *Sam* is added as the right
child of *Pete.* The tree is now

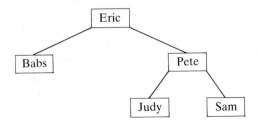

and it is a straightforward matter to add the remaining data and to arrive
at the tree in *Figure 7.4.*

Exercise 7.4

Construct a binary search tree by adding the following items in the
order given below:

26, 13, 51, 43, 28, 6, 62, 17.

7.2 The specification of the binary tree

We shall start our investigation of binary search trees by looking at the
more general case of binary trees in which there is no ordering of items.
The best way to view any binary tree is as follows:

*a binary tree is either empty, or is an item together with two binary
trees.*

This is described in *Fig. 7.5.*

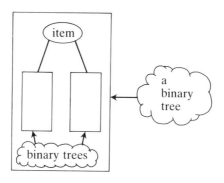

Fig. 7.5 A binary tree

where any of the binary trees shown can be empty. Quite clearly, this is a recursive definition, but it has the considerable advantage of showing the structure of a binary tree. It shows that a binary tree can be decomposed into three parts:

1. an *item* at the top of the tree, usually known as the *root item*;
2. a *left subtree;*
3. a *right subtree.*

Thus, the binary tree shown in *Fig. 7.6* can be decomposed into the item 8 and the left and right subtrees shown in *Figs. 7.7* and *7.8,* respectively.

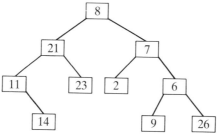

Fig. 7.6 A binary tree example

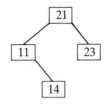

Fig. 7.7 The left subtree

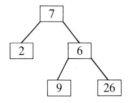

Fig. 7.8 The right subtree

Conversely, a binary tree can be constructed from two existing binary trees and a new root item, by placing the given item in a new root node and joining the existing trees to that root item as left and right subtrees.

The specification of a binary tree will include operations which decompose and construct binary trees in the above manner. It should come as no surprise, therefore, to learn that a binary tree can be specified by the operations listed below.

createtree: requires no source data and returns a new, empty binary tree as its result.

isemptytree: takes an existing binary tree as source data and returns the value *true* when the source tree is empty, or *false* otherwise.

data: takes an existing binary tree as source data and returns as its result the item in the root node when the source tree is not empty, or the message value *empty tree* when the source tree is empty.

left: takes an existing binary tree as source data and returns, as its result, the left subtree.

right: takes an existing binary tree as source data and returns, as its result, the right subtree.

maketree: takes an item and two binary trees as source data and returns a binary tree as its result. The resulting binary tree has the source item as its root node, the first source tree as its left subtree and the second source tree as its right subtree.

For example, assume that *l* and *r* are the binary trees given in *Figs. 7.7* and *7.8*. If *i* represents the item whose value is 8, applying the operation **maketree**(l, i, r) results in the binary tree shown in *Fig. 7.6*.

Exercise 7.5
What are the results of applying the decomposition operators **data**, **left**, and **right** to the following binary tree?

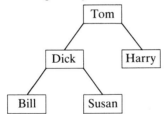

Exercise 7.6
Apply the operation **maketree**(l, i, r) given the following information: *l* and *r* are the binary trees,

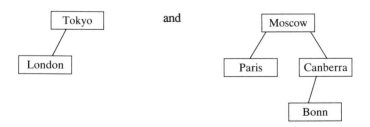

respectively, and *i* is the item *New York*.

With these preliminaries over, we are now in a position to give the formal specification of a binary tree. We shall use the axiomatic approach and produce a generic specification. Here are the *NAME* and *SETS* components of the specification.

NAME
    binary_tree (item)

SETS
    T   the set of binary trees
    I   the set of items
    B   {*true, false*}
    M   {*empty tree*}

Exercise 7.7
Using the *SETS* given above, write down the syntax of the six operations needed in specification of the abstract data type **binary_tree**.

It is instructive to compare the operations which specify a binary tree with those that define a stack. The similarities are striking as *Table 7.1* shows (the numbering indicates similar operations).

Table 7.1   A comparison of stack and binary tree syntax

stack syntax	**binary_tree** syntax
1. **createstack:** $\rightarrow$ S	1. **createtree:** $\rightarrow$ T
2. **isemptystack:** S $\rightarrow$ B	2. **isemptytree:** T $\rightarrow$ B
3. **top:** S $\rightarrow$ I $\cup$ M	3. **data:** T $\rightarrow$ I $\cup$ M
4. **pop:** S $\rightarrow$ S	4a. **left:** T $\rightarrow$ T
	4b **right:** T $\rightarrow$ T
5. **push:** I $\times$ S $\rightarrow$ S	5. **maketree:** T $\times$ I $\times$ T $\rightarrow$ T

Exercise 7.8

Here are the axioms for the semantics of a stack. Deduce the corresponding axioms for a binary tree.

$\forall i \in I, \forall s \in S:$
    isemptystack(**createstack**) = *true*
    isemptystack(**push**(i,s)) = *false*
    top(**createstack**) = *empty stack*
    top(**push**(i,s)) = i
    pop(**createstack**) = **createstack**
    pop(**push**(i,s)) = s

Notice how, in the specification of both stacks and trees, there are operations which construct new objects:

    push and **createstack** in the case of stacks, and
    maketree and **createtree** for binary trees,

and operations which decompose existing objects:

    pop for stacks, **left** and **right** for binary trees.

In both cases the axioms are formed by specifying what happens when a decomposition operator is applied to each construction operator. There are always at least two construction operators: one which brings new empty objects into existence and another which creates a new object from existing object(s).

Exercise 7.9

Write down a recursive definition of an operator, named **number_in**, which can be used to determine the number of items in a binary tree. You will need two rules, one which specifies how the new operator applies to an empty tree, and another which shows how it is applied to a non-empty tree.

7.3   The specification of the binary search tree

A **binary search tree** is a restricted form of binary tree in which the items are *ordered.* That is, there are two relationships, named *comes before* and *comes after,* denoted by < and > respectively, which can be used to place the items in order. The precise meaning of these relationships will depend on the nature of the items: names, for example, are likely to be ordered alphabetically, whereas integers can be placed in ascending or descending order. A binary search tree can be described as follows:

> *a binary search tree is either empty or all the data in the left subtree come before the data in the root, all the data in the right subtree come after the data in the root, and both the left subtree and the right subtree are themselves binary search trees.*

Clearly, a binary search tree can be created for any set of items for which an ordering can be defined.

The purpose of a binary search tree is to facilitate rapid searching of a collection of data items. Therefore, the operations which specify a binary search tree will be somewhat different from those defined for an ordinary binary tree.   In particular, whenever an item is to be added to a binary search tree, it must be inserted in such a way that preserves the specified ordering. For example (*Fig. 7.9*), given the item *i = London,* and the two trees *l* and *r,*

Fig. 7.9 Two binary search trees

then the binary tree given by **maketree**(l, i, r) is (*Fig. 7.10*):

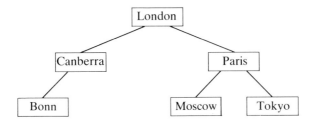

Fig. 7.10 A binary search tree

which is a valid binary search tree. The binary tree obtained from **maketree**(r, i, l) shown in *Fig. 7.11* is, however, *not* a binary search tree.

It would, of course, be possible to redefine **maketree** to produce a message value as its result whenever an attempt was made to construct a tree that was not a binary search tree. This additional complexity is, however, not necessary for our purpose. We shall define a new operation, **insert**, which adds a new item as a leaf node as was done in *Section 7.1*. The specification of **insert** will make use of **maketree**, but in such a way that a binary search tree is guaranteed as the result.

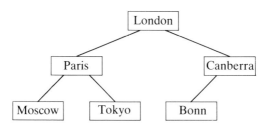

Fig. 7.11 A binary tree which is not a binary search tree

This means that, as far as the specification of a binary search tree is concerned, **maketree** is an operation that is used within the specification but will not be made available in any implementation. Users of an implementation of the abstract data type binary search tree will be able to use the **insert** operation but will not be able to use **maketree**. For this reason, **maketree** is known as a *hidden operation*.

The binary tree decomposition operators **data**, **left**, and **right** have counterparts in the specification of a binary search tree.

We shall add one further operation, **isin**, to the specification of a binary search tree. **isin** takes an item and a binary search tree as source data and returns, as its result, the value *true* when the source item is present in the source tree, or the value *false* otherwise.

For brevity, from now on, we shall refer to a binary search tree as a BSTree.

Exercise 7.10

Write down the *NAME, SETS,* and *SYNTAX* components of the specification of the generic abstract data type BSTree.

There is one feature of all insertion operations (whether for BSTrees or other structures) which must be resolved: what should happen if an attempt is made to insert an item that is already present? Our informal definition of a BSTree does not allow more than one occurrence of an item in any tree. Therefore, we could say that an attempt to insert an item that is already present should result in an message value being returned by **insert**. An alternative, which we shall adopt here to show that there are valid choices to be made, is to simply return the source tree unchanged as the result of such an insertion.

We can now continue the development of the formal specification of a binary search tree by examining the semantics of the eight operations. The first pair of axioms are:

$$\textbf{isemptytree}(\textbf{createtree}) = true \qquad\qquad (T1)$$
$$\textbf{isemptytree}(\textbf{maketree}(l, i, r)) = false \qquad\qquad (T2)$$

where $i \in$ I and $l, r \in$ Bst (the set of BSTrees). It is important to understand fully how the operation **maketree** is used in the second axiom *(T2)*. This axiom states the effect of applying **isemptytree** to a *non-empty* BSTree. One way to ensure that you have a non-empty BSTree is to construct one! The operation **maketree**(l, i, r) will *always* produce a binary tree with at least one item (i) in it, no matter what the state of the trees l and r may be. Another significant point is that the non-empty tree to which **isemptytree** is to be applied must be a binary search tree, and such a tree can always be decomposed into a root item, *i*, and two binary search subtrees, *l* and *r*. It is this item and these two, possibly empty, subtrees which are recombined by **maketree** in axiom *T2*. Hence, the use of **maketree** in axiom *T2* is to show explicity the structure of a non-empty binary search tree; the order of items is automatically preserved by this process.

Exercise 7.11
Write down two axioms (*T3* and *T4*) which relate the meanings of **left**, **maketree**, and **createtree**.

Exercise 7.12
Write down the axioms (*T5, T6*) which specify the semantics for **right**.

Exercise 7.13
Write down the axioms (*T7, T8*) which specify the semantics for **data**.

What you have seen so far should be reminiscent of what we have been doing for other abstract data types. Therefore, we shall now move on to the two operations that are significantly different from those that you have seen so far: **isin** and **insert**.

The operation **isin** requires two items of source data — a data item and an existing BSTree — and returns a Boolean result depending upon whether or not the source data item is in the source tree. There are two cases that we must consider: one when the source tree is empty, the other when the source tree is non-empty. Clearly, when the source tree is empty there are no items in it, and the source data item is, therefore, not present. Hence, the first **isin** axiom is:

$$\textbf{isin}(e, \textbf{createtree}) = \textit{false} \qquad\qquad (T9)$$

where $e \in I$ is the source data item. When the source tree is not empty there are three possibilities:

(i)   the source item is the same as the root item of the source tree;
(ii) the source item may be in the left subtree;
(iii) the source item may be in the right subtree.

In cases (ii) and (iii) it is simply a matter of applying **isin** to the respective subtrees. Hence, the second **isin** axiom is:

isin (e, **maketree**(l, i, r)) = *if* e = i
     then
      true
     else if e < i
      then
       isin(e, l)
      else
       isin(e, r)

which can be written more clearly and succinctly as:

isin (e, **maketree**(l, i, r)) =
    (e = i : *true*   {the item is in the tree}
    | e < i : **isin**(e, l) {check the left subtree}
    | e > i : **isin**(e, r)) {check the right subtree}   *(T10)*

The new notation has the advantage that it exhibits the three possibilities more clearly.

The following example illustrates how it is possible to manipulate the axioms for **isin** in a specific case, and thereby shows you how the axioms achieve our purpose — to specify the operation **isin**.

Suppose that **isin** is to be applied to the BSTree given in *Fig 7.10,* to determine whether or not the item *Moscow* is in the tree. That is, we want to know what the result is of evaluating the expression

  isin(Moscow, t)   *(7.1)*

where *t* represents the BSTree of *Fig. 7.10*. Since the tree *t* is not empty, we can begin by using the definition of the operation **maketree** to rewrite *Expression (7.1)* as

  isin(Moscow, t) = **isin**(Moscow, **maketree**(l, London, r)). *(7.2)*

Here, *l* and *r* are the BSTrees in *Fig. 7.9* (reproduced below).

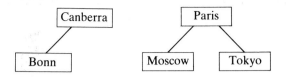

The second **isin** axiom then shows that the right hand side of *Equation (7.2)* is equivalent to the following expression:

(Moscow = London : *true*
| Moscow < London : **isin**(Moscow, l)   *(7.3)*
| Moscow > London : **isin**(Moscow, r)).

As *Moscow > London* is true, *Expression (7.3)* reduces to

isin(Moscow, r).   *(7.4)*

Now *Expression (7.4)* can be rewritten as

isin(Moscow, **maketree**(l', Paris, r'))   *(7.5)*

where *l'* and *r'* are the BSTrees

Moscow   and   Tokyo

respectively. The second **isin** axiom can be applied to *Expression (7.5)* to give:

(Moscow = Paris : *true*
| Moscow < Paris : **isin**(Moscow, l')   *(7.6)*
| Moscow > Paris : **isin**(Moscow, r')).

Because *Moscow < Paris* is true, *Expression (7.6)* reduces to **isin**(Moscow, l'). Again, we can rewrite this as

isin(Moscow, **maketree**(l'', Moscow, r''))   *(7.7)*

where *l''* and *r''* are both empty trees.

The second **isin** axiom gives

(Moscow = Moscow : *true*
l Moscow < Moscow : **isin**(Moscow, l")
l Moscow > Moscow : **isin**(Moscow, r"))

which yields *true,* as expected.

Exercise 7.14

Show that the result of applying the operation **isin** to the item *New York* and the tree, *t,* of *Fig. 7.10* is *false.*

There are two important points to be appreciated about the **isin** axioms:

1. Together they provide a recursive definition of the operation for *any* BSTree and *any* item.

2. Successive application of the second axiom always results in either the value *true* being returned, or **isin** being applied to a *smaller* BSTree. Eventually, therefore, **isin** will either discover that the source item is present in the source tree, or will be applied to an empty tree. In the latter case the first axiom tells us that the required item is not present.

The final pair of axioms define the operation **insert**. We shall begin by investigating what should happen when an item, *e,* is inserted into an empty tree. Clearly the result should be a binary tree with only a root item in it. That is, its left and right subtrees should both be empty. Such a tree can be constructed using **maketree**.

maketree(createtree, e, createtree)

(You should convince yourself that the result is a valid BSTree.) Hence, the first **insert** axiom is given by

insert(e, createtree) = maketree(createtree, e, createtree).   *(T11)*

The second axiom specifies what happens when an item, *e*, is inserted into a non-empty BSTree, *t*. That is, we want to determine the value of **insert**(e, t). This is the same as

insert(e, **maketree**(l, i, r))

where *l* and *r* are binary search trees (since *t* is a BSTree) and $i \in I$. In order to maintain the ordering relationship when the item *e* is added to the source tree, it is clear that if $e < i$ then *e* should be inserted into the left subtree, *l*. However, if $e > i$ then *e* should be inserted into the right subtree, *r*. If it turns out that $e = i$ then we shall return the original tree unchanged. The second **insert** axiom is, therefore,

insert (e, **maketree**(l, i, r)) =
  (e = i : **maketree**(l, i, r)   {return original tree}
  | e < i : **maketree**(**insert**(e, l), i, r) {insert in left subtree} (*T12*)
  | e > i : **maketree**(l, i, **insert**(e, r))) {insert in right subtree}

Exercise 7.15
Explain, in your own words, what is meant by the expression **maketree**(**insert**(e, l), i, r).

Exercise 7.16
If *r* denotes the BSTree containing only the root item *Tokyo,* draw the tree that is the result of evaluating the expression

maketree(**maketree**(**createtree**, New York, **createtree**), Paris, r).

The next example shows how the two **insert** axioms work. Suppose we have a BSTree, *t,* given by

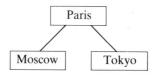

and we want to insert the item *New York* into it. That is, we want the result of evaluating the expression **insert**(New York, t).

We start by using the definition of **maketree** to rewrite the expression as

> **insert**(New York, **maketree**(l, Paris, r))   *(7.8)*

where *l* and *r* are respectively the BSTrees:

| Moscow | and | Tokyo |

Since *New York < Paris,* the second insert axiom, when applied to *Expression (7.8),* gives:

> **maketree**(**insert**(New York, l), Paris, r)   *(7.9)*

which, on applying **maketree** to *l,* can be written as

> **maketree**(**insert**(New York, **maketree**(l', Moscow, r')), Paris, r)
> *(7.10)*

where *l'* and *r'* are both empty trees. The second insert axiom can be applied to **insert**(New York, **maketree**(l', Moscow, r')) in *Expression (7.10)* to give (since *New York > Moscow*)

> **maketree**(**maketree**(createtree, Moscow,
>   **insert**(New York, **createtree**)), Paris, r).   *(7.11)*

It is now appropriate to apply the first **insert** axiom *(T11)* to **insert**(New York, **createtree**) to obtain **maketree**(**createtree**, New York, **createtree**). This means that *Expression (7.11)* now becomes

> **maketree**(**maketree**(createtree, Moscow,
>   **maketree**(createtree, New York, **createtree**), Paris, r).   *(7.12)*

If we denote the tree **maketree**(**createtree**, New York, **createtree**) by t_1 then *Expression (7.12)* becomes

> **maketree**(**maketree**(createtree, Moscow, t_1), Paris, r).   *(7.13)*

The tree **maketree(createtree**, Moscow, t_1) looks like this:

and if we denote it by t_2, *Equation (7.13)* becomes

maketree(t_2, Paris, r)

which can be pictured as shown below.

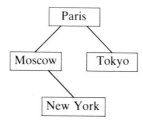

This, of course, is the original tree *t* with the new node inserted.

Exercise 7.17
Show how the insert axioms can build up a BSTree containing only the two items *Oslo* and *Beijing* (inserted in that order).

7.4 Tree traversals

Traversing a tree means visiting every node in the tree in some predetermined way to perform some operation on the data at each node. A common operation of this kind is that of printing out the data value of each node in such a way that the values are printed in order. The recursive nature of a binary search tree yields an elegant solution to this problem.

To see how a traversal algorithm can be built up, look again at *Fig 7.10*, which has been reproduced below.

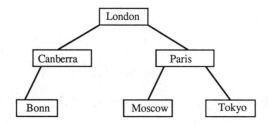

As you have seen before, this tree consists of three parts: a root node, a left subtree and a right subtree. Significantly, *all* the data values in the left subtree come *before* the root node data value, and *all* the data in the right subtree come *after* the data in the root node (check this). This means that a suitable traversal algorithm for printing the data is:

(a)   print out (in order) all the data in the left sub-tree, then
(b)   print out the data in the root node, then
(c)   print out (in order) all the data in the right sub-tree.

This process, applied to *Fig. 7.10*, can be viewed as shown in *Fig. 7.12*.

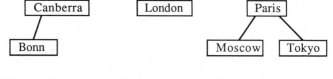

(a) print left subtree   (b) print root   (c) print right subtree

Fig 7.12 Inorder tree traversal

But this is a process that will work for *any* (sub)tree. In particular, if this process is applied to the left subtree of *Fig 7.12(a)*, we would obtain:

(a) print out the left subtree (consisting of the single node containing Bonn);

(b) print out Canberra (from the root node);

(c) print out right subtree (but this causes no output because the right subtree is empty).

Exercise 7.18
Apply the above traversal algorithm to the right subtree of *Fig. 7.10.*

Each repetition of this process yields smaller and smaller subtrees until, eventually, the subtrees are empty and the process stops.

We can amend the above traversal algorithm to take account of the stopping criterion to get

if the tree is *not* empty
then
    print out the left sub-tree
    print out the root node data value
    print out the right sub-tree
if end.

Thus, no action is taken when the (sub-)tree is empty. This process is known as *inorder* traversal.

Exercise 7.19
Apply the inorder tree traversal algorithm to the tree in *Fig. 7.6,* to: (a) determine the order in which the data are printed out, and (b) verify that the recursive application of the process eventually stops.

Such an elegant solution as the inorder traversal algorithm should, of course, be capable of formal specification. This can be achieved by an operation, **inorder**, which takes, as its source data, a binary search tree and returns, as its result, a queue in which elements are the items of the binary search tree in order. That is, **inorder** transforms a binary search tree into a queue in such a way that the items in the queue are in order. Thus, given the sets Bst and Q (of binary search trees and queues, respectively), the syntax of inorder is

    inorder: Bst $\rightarrow$ Q.

inorder is an operation involving two abstract data types. Consequently, its axiomatic specification involves relating the meaning of (some) binary search tree operations and (some) queue operations to one another. The two axioms which, together with the other axioms for a binary search tree and a queue, define **inorder** are

> **inorder(createtree) = createqueue**   *(T13)*
> **inorder(maketree**(l, i, r)) =
>   **appendqueue(addtoqueue**(i, **inorder** (l)), **inorder**(r))   *(T14)*

appendqueue has a syntax of

> **appendqueue**: $Q \times Q \rightarrow Q$

and appends the second mentioned queue on to the tail of the first. Its semantics are specified by the pair of axioms

> **appendqueue(addtoqueue**(i, q), **createqueue**) =
>   **addtoqueue**(i, q)   *(Q7)*

and

> **appendqueue**(p, **addtoqueue**(i, q)) =
>   **addtoqueue**(i, **appendqueue**(p, q)).   *(Q8)*

appendqueue is an abstract data type queue operation which, for brevity and convenience, was not included in the discussion in *Chapter 5*. However, the two axioms involving **appendqueue** are numbered consistently with those of *Fig. 5.3*. We shall not discuss *axioms Q7* and *Q8* any further here, preferring to address the issue of *axioms T13* and *T14*. A reading of the axioms indicates their efficacy in capturing the desired semantics. In particular, *axiom T14* expresses the essence of inorder traversal. That is, the inorder traversal of a binary tree consists of the inorder traversal of its left subtree, followed by the root and, finally, followed by the inorder traversal of its right subtree.

By way of illustration, consider the application of **inorder** to the binary tree of *Fig. 7.13* where the items are integers.

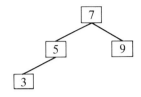

Fig. 7.13 A binary search tree

Axiom T14 can be used to define the result in the following (familiar) fashion: the application of **inorder** is written in the form of the axiom

    inorder(**maketree**(a, 7, b))
       = **appendqueue**(**addtoqueue**(7, **inorder**(a)), **inorder**(b)).   *(7.14)*

Hence, we need to evaluate

    inorder(a)   *(7.15)*

and

    inorder(b).   *(7.16)*

Dealing with (7.15), we write

    inorder(**maketree**(a', 5, **createtree**)) =
       appendqueue(**addtoqueue**(5, **inorder**(a')), **inorder**(**createtree**))
                                               (7.17)

where *a'* is the binary tree with a root containing 3 and empty subtrees. Thus, we next need to evaluate

    inorder(a')   *(7.18)*

and

    inorder(**createtree**).   *(7.19)*

Expression (7.19) evaluates to **createqueue**, by *axiom T13,* and *Expression (7.18)* can be written as

inorder(maketree(createtree, 3, createtree) =
  appendqueue(addtoqueue(3, inorder(createtree)),
    inorder(createtree)) (7.20)

which evaluates to the queue <3>. Using this *(7.17)* then evaluates to
the queue <3, 5>. *Expression (7.16)* will evaluate, in a similar fashion,
to the queue <7>, giving the right hand side of *(7.14)* in the form

  appendqueue(addtoqueue(7, <3, 5>), <9>), (7.21)

that is, the queue <3, 5, 7, 9>.

Exercise 7.20
Here is a different traversal algorithm. In what order are the data of
Fig. 7.10 printed out?

  if the tree is not empty
  then
    print out the root node data value
    print out the left subtree
    print out the right subtree
  if end

The algorithm given in *Exercise 7.20* is known as a *pre-order*
traversal since it prints out the root node *before* either of the subtrees.

Exercise 7.21
Write down the *post-order* tree traversal algorithm which prints out
the root node data value *after* both subtrees have been printed.

Exercise 7.22
Write down axioms defining the operations **preorder** and **postorder**,
which, in a similar fashion to **inorder**, specify the pre-order and the
post-order traversals of a binary search tree, respectively.

7.5 Implementation of the binary search tree

In this section we shall look at the implementation of the abstract data type binary search tree in MODULA-2. For simplicity, we shall choose the tree items to be of type character. Here is the definition module. You should have no difficulty in matching up the parameters with the syntax of the specification.

```
DEFINITION MODULE  BSTreeOps;
TYPE ItemType: CHAR;
TYPE BSTree;        (* To be defined in the IMPLEMENTATION module *)

PROCEDURE  CreateBSTree (): BSTree;
PROCEDURE  IsEmptyBSTree (t: BSTree): BOOLEAN;
PROCEDURE  LeftTree (t: BSTree): BSTree;
PROCEDURE  RightTree (t: BSTree): BSTree;
PROCEDURE  Data (t: BSTree): ItemType;
PROCEDURE  IsIn (e: ItemType; t: BSTree): BOOLEAN;
PROCEDURE  Insert (e: ItemType; VAR t: BSTree);
PROCEDURE  PrintBSTree (t: BSTree);
END  BSTreeOps.
```

Where ever possible we have implemented the BSTree operations as MODULA-2 function procedures. The important exception is *Insert* which will be discussed in more detail later. In this implementation we have included an additional operation, *PrintBSTree,* to illustrate how all the items in a BSTree can be printed out. As with implementations of other abstract data types, the type of the items is specified in the definition module, whereas the abstract data type is defined in the corresponding implementation module. In the jargon of MODULA-2, therefore, *BSTree* is an opaque type.

Notice that the operation **maketree** is not present — in keeping with **maketree**'s role as a hidden operation.

The most natural method of implementation for trees uses pointers. Thus, we can define a *BSTree* as shown below.

```
TYPE
  BSTree =  POINTER TO BSTreeNode;
  BSTreeNode =  RECORD
                    value: ItemType;
                    left: BSTree;
                    right: BSTree;
                END;
```

The implementation of the procedures *CreateBSTree, Left, Right, Data,* and *IsEmptyBSTree,* are all very straightforward and no new concepts are required; here they are.

```
PROCEDURE CreateBSTree (): BSTree;
BEGIN
    RETURN NIL;              (* an empty tree is denoted by the nil pointer *)
END CreateTree;

PROCEDURE IsEmptyBSTree(t: BSTree): BOOLEAN;
BEGIN
    RETURN (t = NIL);
END IsEmptyBSTree;

PROCEDURE LeftTree (t: BSTree): BSTree;
BEGIN
    IF IsEmptyBSTree(t)
    THEN
        RETURN NIL;
    ELSE
        RETURN t↑.left;
    END;
END Left;

PROCEDURE RightTree (t: BSTree): BSTree;
BEGIN
    IF IsEmptyBSTree(t)
    THEN
        RETURN NIL;
    ELSE
        RETURN t↑.right;
    END;
END Right;

PROCEDURE Data (t: BSTree): ItemType;
BEGIN
    IF IsEmptyBSTree(t)
    THEN
        WriteString('Tree is empty: cannot obtain data.');
        WriteLn;
    ELSE
        RETURN t↑.value;
    END;
END;
```

The implementation of the operation **isin** follows the specification quite closely so we have shown them both side-by-side for comparison.

```
PROCEDURE  IsIn (e: ItemType; t: BSTree):  BOOLEAN;
BEGIN
  IF  IsEmptyBSTree (t)
  THEN  (* First axiom applies *)
    RETURN FALSE;
  ELSE  (* Second axiom applies *)
    IF e = Data(t)                              ( e = i
    THEN                                        :
      RETURN TRUE;                              true
    ELSE                                        |
      IF e < Data(t)                            e < i
      THEN                                      :
        IsIn (e, LeftTree(t));                  isin(e, l)
      ELSE                                      |
        IsIn (e, RightTree(t));                 isin(e, r))
      END;
    END;
  END;
END IsIn;
```

Before we can implement **insert** certain questions must be answered. In the definition module we have already chosen to use a **VAR** parameter for the source tree. Our aim is to perform the insertion operation in such a way that an additional node is appended to the source tree, in the appropriate place. In other words we shall not be producing a totally new tree as the result, merely an updated version of the source tree. Here is the procedure.

```
PROCEDURE  Insert (e: ItemType; VAR t: BSTree);
VAR  temp: BSTree;
BEGIN
  IF  IsEmptyBSTree(t)
  THEN     (* First axiom applies *)
    Allocate(temp, SIZE(BSTreeNode));
    WITH temp↑ DO
      data := e;
      left := NIL;  right := NIL;
    END;
    t := temp;
  ELSE     (* second axiom applies *)
    IF  e < Data(t)
    THEN
      Insert (e, t↑.left);
    ELSE
      Insert (e, t↑.right);
    END;
  END;
END Insert;
```

The most significant point to understand about this procedure is what happens when it is called recursively. Each recursive call is provided with a value for *t,* a pointer to the root node of a subtree. Eventually the situation will occur when the value of *t* is **NIL**, indicating an empty tree. At this stage, that part of the procedure dealing with the actions of the first axiom will be executed, and a new node will be created with the appropriate values entered into it. On exit from the procedure, having created the new node, the value of the formal parameter *t* will be a pointer to the new node. Since the procedure uses a **VAR** parameter, the value of the corresponding actual parameter will be the same pointer value. This time, depending on which of the two recursive calls was made, the actual parameter is either *t* ⇑. *left* or *t* ⇑. *right,* which refer to the pointer fields of a leaf node. Hence, the new node has been linked to an existing leaf node of the source tree, as required.

The final operation to be considered is *PrintBSTree.* We shall implement the inorder traversal algorithm

```
PROCEDURE PrintBSTree (t: BSTree);

    PROCEDURE PrintItem (e: ItemType);
    BEGIN
        Write(e);
        WriteLn;
    END PrintItem;

BEGIN
    IF NOT IsEmptyBSTree(t)
    THEN
        PrintBSTree (LeftTree(t));
        PrintItem (Data(e));
        PrintBSTree (RightTree(t));
    END;
END PrintBSTree;
```

We have separated out the printing of individual items into the procedure *PrintItem* so that it will be easy to modify the routine whenever the type of item is changed.

Exercise 7.23

Trace the procedure *PrintBSTree* when it is applied to the BSTree given in *Figure 7.4.* (Solution not given).

7.6 Summary of chapter

This chapter has examined trees and introduced a great deal of terminology associated with them:

node branch leaf root subtree child parent level

A tree in which the maximum number of children of any one node is two is a **binary tree**, and is defined recursively as:

a binary tree is either empty or is an item together with two binary trees.

Six operations were defined on binary trees:

createtree, isemptytree, data, left, right, and **maketree.**

If the data to be stored in a binary tree can be ordered it is possible to create a **binary search tree**.

A binary search tree is either empty or all the data in the left subtree come before the data in the root node, all the data in the right subtree come after the data in the root node, and both the right and left subtrees are themselves binary search trees.

The formal specification of the abstract data type binary search tree was given using the axiomatic approach. There are 8 operations defined for a binary search tree:

createtree, isemptytree, data, left, right, maketree, isin, and **insert**

Visiting each node in a tree in a pre-determined order is known as *traversing* the tree. To print out the data in a binary search tree in order requires an *inorder* tree traversal. Two other kinds of tree traversal were exhibited: *pre-order* and *post-order traversals*. A formal specification was given for inorder traversal.

The chapter concluded with an implementation of the abstract data type binary search tree in MODULA-2. The implementation was based on a linked representation.

8 Case studies

In previous chapters, we have discussed the application of abstract data types to some very well known and well understood situations. In this chapter, we shall examine a small number of case studies to see how abstract data types and information hiding can be used in practice. The case studies are small when compared to real problems but they are sufficiently different from what you have seen so far for you to gain a realistic assessment of the approach.

Each case study is divided into three subsections, the first of which informally describes the application for which the abstract data type is required. This will enable us to discover, still in an informal way, the characteristics of the new abstract data type. Once we have an intuitive feel for the new object we shall be able to specify it formally in the next subsection. A third subsection will discuss a representation of the abstract data type, and the final subsection will look at some of the details of implementation in either Pascal, MODULA-2, or Ada.

8.1 Case study 1: a directory

8.1.1 The application

A company wants to keep an up-to-date internal telephone directory on a computer. The problem has already been analysed, and a requirements list has been drawn up showing that the following operations must be carried out on the directory if it is to meet all predictable demands:

(i) adding an employee's name and telephone number;

(ii) removing an employee's name and associated telephone number;

(iii) retrieving an employee's telephone number, given the employee's name;

(iv) displaying the whole directory in alphabetic order of employee name;

(v) creating a new, empty telephone directory.

The above descriptions are sufficient to indicate the scope of the application, but we will need more detailed information about the individual operations.

Adding a new employee name and telephone number appears, at first sight, to be a straightforward operation. But what should happen if there already exists an entry in the directory for an employee with that same name? We shall take the view that duplicate entries in the directory are not allowed and, therefore, this operation should not add duplicates, but should output a suitable message whenever such an attempt is made.

Another kind of difficulty occurs with the retrieval and removal operations. What should happen if an attempt is made to retrieve or remove the telephone number of an employee for whom no entry has yet been made in the directory? Again we shall require the operations to output a suitable message. This discussion suggests that we could give application programmers additional flexibility by providing an operation which checks whether an employee with a specified name already exists in the directory.

The ability to change an employee's telephone number can be provided by combining the effects of two operations — first remove the existing entry, then add a new entry containing the revised information. In effect we are defining a set of *atomic* operations, out of which more complex procedures can be built.

The display operation is the least well defined of all the operations, and ought to be clarified in discussions with users to determine precisely what is required. For our purposes we shall assume that the contents of a directory should be printed out in alphabetical order of employee name, in a suitable format (which we shall not pursue further), at a terminal.

We can summarize these points by outlining the six required operations.

add: takes an employee name, a telephone number, and a directory as source data, and returns a directory with the employee data inserted. If the employee name already exists in the directory, a suitable message is returned.

remove: takes an employee name and a directory as source data and, if that employee name exists in the directory, removes the name and the associated telephone number from the directory and returns the modified directory as its result. If the employee name is not in the directory a suitable message is given.

retrieve: takes an employee name and a directory as source data and, if that employee name exists in the directory, returns the associated telephone number. If the employee name is not in the directory a suitable message is returned.

display: takes a directory as its source data and causes a listing (in alphabetical order of employee name) to be printed at a terminal.

createdirectory: takes no source data and returns a new, empty directory as its result.

isindirectory: takes an employee name and a directory as source data and returns the value *true* if the employee name exists in the directory, and the value *false* otherwise.

8.1.2   The specification

A directory is a collection of data about employees, and we shall draw up a specification for a particular application (i.e. not a generic specification). The *NAME, SETS,* and *SYNTAX* components of the specification are given below.

NAME
  directory(employee)

SETS
  D   the set of all directories
  N   the set of employee names
  T   the set of telephone numbers
  B   {*true, false* }
  M   {*employee not in directory, employee already in directory* }
  TL   the set of terminal listings

SYNTAX

createdirectory: $\rightarrow$ D
isindirectory: N $\times$ D $\rightarrow$ B
retrieve: N $\times$ D $\rightarrow$ T $\cup$ M
add: N $\times$ T $\times$ D $\rightarrow$ D $\cup$ M
remove: N $\times$ D $\rightarrow$ D $\cup$ M
display: D $\rightarrow$ TL

SEMANTICS

We shall use the axiomatic approach. Therefore, we shall need a (hidden) construction operator, which we shall name **makedirectory**, with the following property. It will take, as source data, an employee name, a telephone number, and a directory and will produce as its result a new directory consisting of the source directory with a new directory item added at its front. The new directory item will contain the employee name and telephone number. This operation is guaranteed to produce a non-empty directory as its result. As with other abstract data types we shall use the operation **makedirectory** either when we want to be sure that a given directory is non-empty, or to show the structure of the directory (it consists of a first item and a (sub) directory). A typical use of this operator is **makedirectory**(n, t, d), where n and t are from the sets N and T, respectively, and d is a directory. Pictorially, we can represent a directory as shown in *Fig. 8.1,* in which the items are ordered alphabetically by employee name.

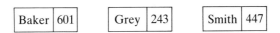

| Baker | 601 | | Grey | 243 | | Smith | 447 |

Fig. 8.1 A directory

The operation **makedirectory** should be formally defined. Here is one way to tackle this problem. We know that two lists can be joined together by the operation **concatenate** so, by using a list to model a directory, we can join two directories together by the same means. However, **makedirectory** takes an item and adds it to the front of an existing directory. Therefore, if we have an operation, **convert**, which takes an employee name and a phone number as source data, and produces a directory containing one item, then we can define

makedirectory(n, t, d) = **concatenate**(**convert**(n,t), d).

Thus we have defined **makedirectory** in terms of two simpler operations. One can continue in the same way by arguing that **convert** and **concatenate** should now be formally defined. However, it is commonly accepted that **convert** and **concatenate** are so well known as to require no further elaboration.

Turning now to the axioms, if $e \in N$, i.e. an employee name, then the axioms for **isin** are:

isindirectory   (e, **createdirectory**) = *false*   *(D1)*

and

isindirectory   (e, **makedirectory**(n, t, d)) =
    (e = n :   *true*   {name is in the directory}
    | e < n : *false*   {name is not in directory}   *(D2)*
    | e > n : **isindirectory**(e, d))   {check the sub-directory d}

In *axiom D2* we are making use of the fact that the items in the directory are ordered alphabetically by employee name and, therefore, the employee name in the first item always comes before all the other names in the directory. The operators $<$ and $>$ are the usual relational operators on names. The use of the operation **makedirectory** as a parameter to **isindirectory** is as a deconstructor, and says that, in the case of a non-empty directory, the directory consists of a first item (holding two pieces of data n and t), at the front of a sub-directory d. Of course, the sub-directory d could be empty. When this happens, *axiom D1* would be applied.

Two axioms are required to define the **add** operation ($p \in T$).

add (e, p, **createdirectory**) = **makedirectory** (e, p, **createdirectory**)
                                                             (D3)
add (e, p, **makedirectory**(n, t, d)) =
    (e = n : *employee name already in directory*
    | e < n : **makedirectory**(e, p, **makedirectory**(n, t, d))   *(D4)*
    | e > n : **makedirectory**(n, t, **add**(e, p, d)))

Axiom D3 simply states that when adding an item to an empty directory it must end up being the first item and that, of course, is precisely what **makedirectory** does! *Axiom D4* is just a little more complicated. The case when the item to be added has the same employee name as the item at the front of the sub-directory is straightforward: the message *employee name already in directory* should be returned, to avoid adding duplicate employee names. When the name to be added, *e,* comes before the name at the front of the current sub-directory, (i.e. $e < n$), the new name should be added to the front of the sub-directory. The final case, $e > n$, means that the new name should be placed in the directory somewhere after the first item, (n, t), so it must be **added** to the sub-directory *d*.

Moving on to the specification of the semantics of the **retrieve** operation, we have, for all $e, n \in$ N; $t \in$ T; and $d \in$ D:

retrieve(e, **createdirectory**) = *employee not in directory*   *(D5)*

retrieve (e, **makedirectory** (n, t, d)) =
  (e = n : t   {name found — return telephone number}
  | e < n : *employee not in directory*   *(D6)*
  | e > n : **retrieve**(e, d))   {examine remainder of directory}.

Exercise 8.1
The axioms for **remove** are similar to those for **retrieve**; what are they?

The formal specification of **directory** will be complete once we have examined the **display** operation. This operation requires that the complete contents of a directory are printed out, in order, at a terminal. A full specification of this operation would take into account the format for printing out the data. Such detail can often be very involved and may include characteristics of the terminal device. Here we do not want to get involved in such detail, so we shall assume the existence of an operation, named **displayitem,** which will be responsible for displaying the contents of a single item in the required format for a specific terminal type. This will enable us to capture the essential properties of the **display** operation. Here are the axioms for **display**. For all $n \in$ N, $t \in$ T, and $d \in$ D:

display(**createdirectory**) = **createdirectory**,   *(D9)*

that is, do nothing if the directory is empty.

display(**makedirectory**(n, t, d)) = **displayitem**(n, t); **display**(d)   *(D10)*

Thus, take the first item (*n, t*) from the directory, display it, and then perform the same operation on the remaining directory, *d*. We have used the semicolon to indicate the sequence of actions. You should compare this technique with what we did with the inorder tree traversal in *Section 7.5*.

8.1.3   The representation

We shall represent a directory as a linked sequence of records (see *Fig. 8.2*). This representation includes a dummy item so that an empty sequence can be treated in the same way as any other sequence.

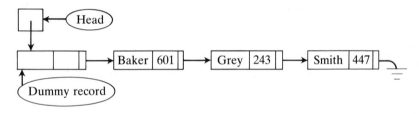

Figure 8.2   A representation of the directory

8.1.4   The implementation

The directory will be implemented in Pascal and we shall concentrate on the bodies of the procedures which implement a selection of the directory operations assuming that, in any actual implementation, the principles of information hiding will be rigourously followed.

The necessary **type** definitions are shown below.

```
type
  Name =  array[ 1 .. 20 ] of Char;
  PhoneNo = Integer;
  Link =  ↑NodeRecord;
  NodeRecord =  record
                  EmployeeName: Name;
                  Telephone: PhoneNo;
                  Next: Link
                end;
  Directory = Link;
```

The procedure *CreateDirectory* is similar to others of the same type.

```
procedure CreateDirectory( var d: Directory);
begin
  New(d);
  with d↑ do
  begin
    EmployeeName := '';
    Telephone := 0;
    Next := nil
  end
end {CreateDirectory};
```

To add a new item means finding the correct insertion point in the sequence and then linking in the new item. To find the insertion point requires two pointer variables, *Current* and *Previous,* to indicate the two existing items between which the new item is to go. *Fig. 8.3* shows how the link fields must be altered to incorporate the item containing the name *Rogers* into the sequence shown in *Fig. 8.2*.

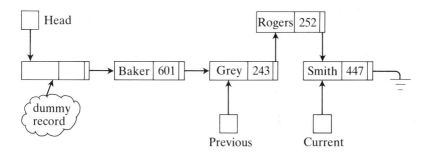

Fig. 8.3 Adding an item

```
procedure Add (n: Name; t: PhoneNo;
              var d: directory);
var  Previous, Current, Temporary: Link;
     Found: Boolean;        {true if insert position found or at end of sequence}
begin
  Previous := d;                    {points to dummy item}
  Current := Previous↑.Next;   {points to first item or is nil if directory is empty}
  Found := (Current = nil );
  while not Found do                    {search through list}
    if  n < Current ↑.EmployeeName
    then  Found := true
    else
      begin                          {move on to next item}
        Previous := Current;
        Current := Current↑.Next;
        Found := (Current = nil )
      end;
  if Previous ↑.EmployeeName = n
  then  Writeln('Employee already in directory')
  else
    begin            {create new record and add into sequence}
      New(Temporary);
      with Temporary↑ do
        begin
          EmployeeName := n;
          Telephone := t;
          Next := Current;
        end;
        Previous ↑.Next := Temporary
    end
end {Add};
```

Fig. 8.4 illustrates what must happen when an item is removed from a
directory: *Current* points to the item to be removed and *Previous*
points to the record whose *Next* field must be updated.

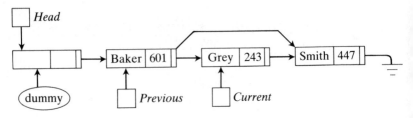

Fig. 8.4 Removing an item

```
procedure Remove(n: Name;  var d: directory);
var
    Previous, Current: Link;
    Found, EndOfList: Boolean;    {true if item found / true at end of sequence}
begin
    Previous := d;                {points to dummy item}
    Current := Previous ↑.Next;   {points to first item/is nil if directory empty}
    Found := false;
    EndOfList := (Current = nil );
    while (not Found) and (not EndOfList) do  {search through list}
        if  n = Current ↑.EmployeeName
        then
            Found := true
        else
            begin                 {move on to next item}
                Previous := Current;
                Current := Current ↑.Next;
                EndOfList := (Current = nil )
            end;
    if Found
    then
        Previous ↑.Next := Current↑. Next
    else
        Writeln ('Employee not in directory')
end {Remove};
```

Exercise 8.2
Write down the Pascal procedures which implement the operations
retrieve, display, and **isindirectory.**

Exercise 8.3
Design and implement a Pascal program to update a directory when an employee's telephone number is changed.

8.2 Case Study 2: a file editor

8.2.1 The application

This case study is based on serial file processing of the kind supported in Pascal. That is, a file is a collection of records that can be accessed serially, and there are a number of permitted operations that the programmer can perform on the file. A file is a *sequence* of records, and the kinds of operations that can be performed include: inserting a new record into a file, deleting a record from a file, and replacing one record by another. In this discussion, the precise structure of a record will be immaterial. The manner in which records are inserted in, and deleted from, a file depends on the concept of a **current record**. The user views a file as a sequence of records, one of which (the *current record*) is currently being examined. This is very much like the *window* concept in a Pascal file. At any point in time there is precisely one record in the window which the programmer can process, and to access other records the window must be moved. However, unlike Pascal files, we shall allow insertion and deletion in the middle of a file. Unless the current record is either the first or the last record in a file there will be records in the file which come before the current record, and records which come after the current record. The user is able to traverse the file by altering the current record. This is achieved by moving to an adjacent record, either forwards or backwards in the file. The operations on a file are defined below.

createfile: creates an empty file.

isemptyfile: determines whether or not the file is empty.

insert: inserts a new record into the file immediately after the current record and makes the new record the current record.

delete: deletes the current record; the previous record becomes the current record.

forward: makes the next record the current record; if the current record is the last record no action is taken.

backward: makes the previous record the current record; if the current record is the first record no action is taken.

replace: replaces the current record by a new record.

current: returns the current record.

Any attempt to access a record in an empty file results in a message being reported.

8.2.2 The specification

Here we shall use the constructive approach. To do so, we need an appropriate underlying model. This abstract data type has two characteristics which have to be modelled: the *sequence* of records and the identification of the current record (that is, denoting which is the current record). It is this latter characteristic which makes this case study different from the other abstract data types discussed in this text.

There are several ways to model the file. The one we have chosen is not, perhaps, the most obvious but does employ many of the concepts mentioned in this book. Our model will use *three* lists.

the left list	the middle list	the right list

Thus, any file can be split into three parts. The middle part is a list containing just one record — the current record. To the left of the *current list* is the *left list* containing the sequence of records which comes *before* the current record. To the right of the *current list* is the *right list* containing the sequence of records which comes *after* the current record. Concatenating the left list, the current list, and the right list (in that order) results in a file. Of course, any or all of the three lists can be empty. When the current list is empty the other two lists must also be empty, and this signifies an empty list.

The fact that we are working with a composite model means that we shall need some additional notation to describe a file in the formal specification. Let L be the set of all lists and let a, b, and c be instances of lists (i.e. $a, b, c \in L$), then the *triple* $< a, b, c >$ is an instance of a file (i.e. $< a, b, c > \in F$, where F is the set of all files). Recall that a triple is ordered. A similar notation was used for lists.

We are now in a position to be able to present the formal specification using the constructive approach. It is a generic specification in that the items in a file are not specified (we have renamed the items as *records* because that is the conventional term for items, which can be composite objects, held in a file).

NAME
file(records)

SETS

R	the set of all records
L	the set of all lists
M	the set consisting of the single message *error*
B	{*true, false*}
F	L × L × L

(The definition of the set F implies that an element of this set is composed of three lists.)

SYNTAX

createfile: → F

isemptyfile: F → B

insert: F × R → F

delete: F → F ∪ M

forward: F → F ∪ M

backward: F → F ∪ M

replace: F × R → F ∪ M

current: F → R

SEMANTICS

The underlying model consists of three lists: $l, c, r \in$ L; and $i \in$ R. In what follows *all* pre-conditions are *true* so only the post-conditions will be exhibited (this follows the approach of Jones, 1980).

post-**createfile**(f) ::= f = <**createlist, createlist, createlist** >   *(F1)*

post-**isemptyfile**(< l, c, r>; b) ::= b = (c = **createlist**)   *(F2)*

post-**insert**(< l, c, r>, i; f) ::= f = < **concatenate**(l, c), **make**(i), r >

(F3)

post-**forward**(< l, c, r>; f) ::=
    if c = **createlist**
    then
      f = *error*
    else if r = **createlist**
      then
        f = < l, c, r >   *(F4)*
      else
        f = < **concatenate**(l, c), **make** (**first**(r)), **trailer**(r) >

post-**backward**(< l, c, r>; f) ::=
     if c = **createlist**
     then
       f = *error*
     else if l = **createlist**
        then
         f = < l, c, r >   *(F5)*
        else
         f = < **leader**(l), **make** (**last**(l)), **concatenate**(c, r) >

post-**replace**(< l, c, r>, i; f) ::= *if* c = **createlist**
                            then
                              f = *error*
                            else
                              f = < l, **make**(i), r >   *(F6)*

post-**delete**(< l, c, r>; f) ::= *if* (c = **createlist**) ∨ (l = **createlist**)
                      then
                        f = *error*
                      else
                      f = < **leader**(l), **make** (**last**(l)), r >   *(F7)*

(In the **delete** operation it is the *previous* record which becomes the current record, and this accounts for the condition after the *if*.)

post-**current**(< l, c, r>; i) ::= i = c   *(F8)*

Exercise 8.4
Repeat the constructive specification of the semantics of the file abstract data type, this time with an underlying model that employs *two* lists.

8.2.3 The representation

A file is a dynamic structure and therefore a linked representation is appropriate. There is little advantage in representing a file as three separate lists. Instead, the file will be represented as a single list together with a pointer to the current record. An empty list will be represented by the current pointer having the value **nil** (*Fig. 8.5*).

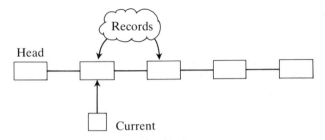

Fig. 8.5 Representation of a file

To facilitate moving the current pointer both forwards and backwards through the file a two-way linked list can be used (*Fig. 8.6*).

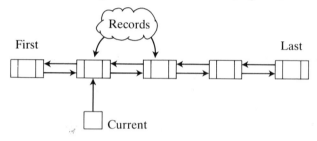

Fig. 8.6 A two-way linked list

Each item has both a left link (pointing to its predecessor) and a right link (pointing to its successor). Since the item of attention is always indicated by the *current* pointer there is no need to indicate the first or last item of the list explicitly.

There is a need, in the **delete** operation to know when the current record is the first record in the list in order to detect an error situation. This information is easily obtained by looking at the left pointing links: the first record has a **nil** left link. Similarly, the **forward** operation needs to know when the last record is the current record, and this is indicated by the last record having a **nil** right link.

8.2.4 *The implementation*

Here is the implementation of the abstract data type **file** in MODULA-2. First, the definition module:

```
DEFINITION MODULE FileOps;
TYPE
   ItemType = RECORD
                  Letter: CHAR;          (* We have chosen a simple form *)
                  Number: INTEGER        (* for file records *)
              END;
TYPE
   File;                    (* type defined in IMPLEMENTATION module *)

PROCEDURE CreateFile ( VAR f: File);
PROCEDURE IsEmptyFile ( f: File): BOOLEAN;
PROCEDURE Insert ( i: ItemType; VAR f: File);
PROCEDURE Delete ( VAR f: File);
PROCEDURE Forward ( VAR f: File);
PROCEDURE Backward ( VAR f: File);
PROCEDURE Replace ( i: ItemType; VAR f: File);
PROCEDURE Display ( f: File);

END FileOps.
```

Here is a partial implementation (see *Exercise 8.5*):

```
IMPLEMENTATION MODULE FileOps;
FROM InOut IMPORT Write, WriteString, WriteLn, WriteInt;
FROM System IMPORT Allocate, Deallocate;
TYPE
   File = POINTER TO FileRecord;
   FileRecord = RECORD
                   Data: ItemType;
                   Left: File;
                   Right: File
                END;

PROCEDURE CreateFile ( VAR f: File);
BEGIN
   f := NIL
END CreateFile;

PROCEDURE Insert (i: ItemType; VAR f: File);
VAR
   Temp: File;
BEGIN
   Allocate(Temp, SIZE (FileRecord));
   Temp ↑.Data := i;
   IF f = NIL
   THEN
      Temp ↑.Left := NIL;
      Temp ↑.Right := NIL
   ELSE
      Temp ↑.Left := f;
```

```
          Temp ↑.Right := f↑.Right;
          IF f↑.Right <> NIL
          THEN
              f↑.Right↑.Left := Temp
          END;
          f↑.Right := Temp
       END
       f := Temp
    END Insert;

    PROCEDURE Forward ( VAR f: File);
    BEGIN
       IF f = NIL
       THEN
         WriteString("Empty File."); WriteLn
       ELSE
          IF f↑.Right = NIL
          THEN
            (* no action *)
          ELSE
             f := f ↑.Right
          END
       END
    END Forward;

    PROCEDURE Replace (i: ItemType;  VAR f: File);
    BEGIN
       IF f = NIL
       THEN
          WriteString("Empty File."); WriteLn
       ELSE
          f↑.Data := i
       END
    END Replace;

    PROCEDURE Display (f: File);
    BEGIN
       IF f = NIL
       THEN
          WriteString('Empty File.'); WriteLn
       ELSE
          Write (f↑.Data.Letter);              (* These are implementation-specific *)
          WriteInt (f ↑.Data.Number);   (* and will change if ItemType changes *)
          WriteLn
       END
    END Display;

    END FileOps.
```

Exercise 8.5

Complete the implementation of the file abstract data type by writing the procedures *IsEmptyFile*, *Delete*, and *Backward*.

8.3 Case study 3: a database application

8.3.1 The application

A *relational database* is a collection of *relations* on which certain *relational operations* can be performed. Here are some typical relations for a suppliers and products database.

Products	Product name	Product number	Description
	Nail	0864	Oval
	Screw	9573	Countersunk
	Nut	2736	Hexagonal
	Bolt	9283	Long
	Washer	6453	Large

Supplier	Supplier code	Supplier name	Supplier address	Credit period
	S39	AJ Smith	Luton	2
	E42	GGJones	Norwich	3
	N08	Pepper & Co	Chester	2

Prices	Product Number	Supplier Code	Cost
	0864	E42	1.03
	2736	E42	1.65
	6453	N08	0.60
	6453	S39	0.58
	9283	S39	7.41
	9573	E42	3.22
	9573	S39	3.50

From the point of view of the *database administrator,* the person who is responsible for the correct maintenance of the database, the structure of the database is important information. The structure of a database includes such information as:

the names of the relations currently in the database (for example, Supplier, Product);

the structure of each relation. (For example, each row in the Prices relation has three items of data named Product number, Supplier code and Cost. These names are the names of the relation's *attributes.*);

the collection of information about the structure of a database is called a *schema* (see Date, 1986). It is the database administrator's job to maintain the schema by adding and deleting relations in the database.

For the purposes of this case study we shall investigate the operations that define a relational database from the point of view of maintaining a schema. Relations are identified by name (for example, Products), and can be added, deleted, and retrieved from a schema. In addition it will be possible to determine whether or not a specific relation is currently in the database schema. Here are the informal specifications of these operations.

createschema: takes no source data and returns a new, empty schema as its result.

isemptyschema: takes a schema as source data and returns true if the schema is empty, and false otherwise.

addrelation: takes a schema, a relation name, and a relation description as source data, and returns the schema with the relation added to it.

deleterelation: takes a schema and a relation name and returns a schema with the relation deleted.

getrelation: takes a schema and a relation name as source data and returns the relation description as its result.

isinschema: takes a schema and a relation name and returns true if the relation is in the schema and false otherwise.

8.3.2   The specification

The following specification uses the axiomatic approach. In many ways, the schema abstract data type is similar to the directory in that similar kinds of operations are defined, but there are some significant differences. In particular, there is no requirement that the relations are held in any specific order in the schema. This fact is reflected in the specification of the operations **addrelation, deleterelation,** and **getrelation**.

NAME
    schema(relation)

SETS
S	the set of schemas
R	the set of relation descriptions (each a list of attribute names)
N	the set of relation names
M	{*relation already in schema, relation not in schema* }
B	{*true, false*}

SYNTAX
    createschema: $\rightarrow$ S
    isemptyschema: S $\rightarrow$ B
    isinschema: N $\times$ S $\rightarrow$ B
    addrelation: N $\times$ R $\times$ S $\rightarrow$ S $\cup$ M
    deleterelation: N $\times$ S $\rightarrow$ S
    getrelation: N $\times$ S $\rightarrow$ R $\cup$ M

SEMANTICS
The specification of the semantics uses a hidden operator **makeschema** which constructs a new schema from an existing schema by adding information about a relation to the original schema. The notation **makeschema**(n, r, s) means: return a schema made up from the existing schema, *s,* plus the relation name, *n,* and the relation description, *r.* A schema is a *set* of objects, and the objects are ordered pairs of relation names and relation descriptions *(n, r).* Incidentally, since a schema is a set, adding a relation name–description pair which already exists in the set results in the same set since, by the definition of a set, a set element can occur only once in a set. Hence,

$\forall$ m, n $\in$ N; r $\in$ R; s $\in$ S:

isemptyschema(**createschema**) = *true*   (S1)

isemptyschema(**makeschema**(n, r, s)) = *false*   (S2)

isinschema(m, **createschema**) = *false*   (S3)

isinschema(m, **makeschema**(n, r, s)) = *if* n = m
$\qquad\qquad$ *then*
$\qquad\qquad\qquad$ *true*
$\qquad\qquad$ *else*
$\qquad\qquad\qquad$ **isinschema**(m, s)   (S4)

Since the schema is not ordered *axiom S4* states that the relation, *m* , is either the same as relation *n* in the schema, or the rest of the schema (i.e. excluding relation *n*) must be examined.

addrelation(n, r, s) = *if* **isinschema**(n, r, s)
$\qquad\qquad$ *then*
$\qquad\qquad\qquad$ *relation already in schema*
$\qquad\qquad$ *else*
$\qquad\qquad\qquad$ **makeschema**(n, r, s)   (S5)

There is only one axiom for **addrelation** because its function is the same as **makeschema** but with a check on whether the relation is already in the schema.

deleterelation(m, **createschema**) = **createschema**   (S6)

deleterelation(m, **makeschema**(n, r, s)) =
$\qquad$ *if* m = n
$\qquad$ *then*
$\qquad\qquad$ **deleterelation**(m, s)
$\qquad$ *else*
$\qquad\qquad$ **makeschema**(n, r, **deleterelation**(m, s))   (S7)

Axiom S7 is based on the idea of working through *all* the relations in a schema, each time picking out a relation, examining it and then putting it back whenever it is *not* the one to be deleted. The operation **makeschema**(n, r, s) picks out the item *(n, r)* leaving the remainder in the subset *s*. The equation

> **deleterelation**(m, **makeschema**(n, r, s) =
>> **makeschema**(n, r, **deleterelation**(m, s))

causes the operation **deleterelation** to be applied to successively smaller subsets of the original schema (eventually terminating with the empty schema whereupon *axiom S6* applies). At some stage during the recursion the process will pick out the item to be deleted ($m = n$), if it exists. When this happens, we do not want to put it back into the schema so we omit the use of **makeschema**. To convince yourself that these axioms do work correctly see what happens when the relation *b* is to be deleted from the schema { *a, b, c* }. You may also like to compare the above specification with that given in Gehani,1986.

> **getrelation**(m, **createschema**) = *relation not in schema*   *(S8)*

> **getrelation**(m, **makeschema**(n, r, s)) = *if* m = n
>> *then*
>>> r
>>
>> *else*
>>> **getrelation**(m, s)   *(S9)*

Axiom S9 is somewhat similar to *axiom S7* but is less complicated. Here we are returning a single relation description (as opposed to a complete schema in *axiom 8*). The axiom effectively looks at each individual relation and, if it is not the desired relation, excludes it from the set of relations still to be examined.

Exercise 8.6

A simple **set** is an abstract data type that is defined by the following operations:

createset: creates a new, empty set;

in: returns *true* if the source data item is in the source set, *false* otherwise;

add: adds an item to a set;

delete: removes an item from a set.

Write down the axiomatic specification of the semantics of simple **set**.

8.3.3   The representation

Since a schema is a collection of relations a simple linked list will suffice for the representation. Each element of the list will represent a relation. For each relation we need to keep a record of its name together with a description of its attributes. *Fig. 8.7* shows a picture of our representation.

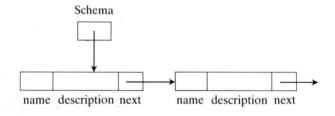

Fig. 8.7   The representation of a schema

8.3.4   The implementation

For this implementation we shall use Ada. However, since a simple linked list has been dealt with earlier, we shall present only the package specification.

```
package  SchemaOps  is
    subtype  RelName  is String (1 . . 20);
    subtype  RelDescription is String (1 . . 30);

    procedure CreateSchema ( S: out  Schema);
    procedure IsEmptySchema ( S: in  Schema): Boolean;
    procedure IsInSchema ( N:  in  RelName;  S:  in out Schema): Boolean;
    procedure Addrelation ( N:  in  RelName;  D:  in RelDescription;
                            S:   in out  Schema);
    procedure DeleteRelation ( N:  in  RelName;  S:  in out Schema);
    procedure GetRelation ( N:  in  RelName;  S:  in  Schema;
                            D:   in out  RelDescription);

private
    type  Link is access  Relation;
    type  Relation  is
            record
                Name: RelName;
                Description: RelDescription
            end record;
    type  Schema  is  Link;
end SchemaOps;
```

9 Object-oriented program design

9.1. Introduction

Throughout this book our emphasis has been on the formal **specification** of abstract data types. Our concern has also been to ensure a faithful implementation of a formally specified abstract data type. The case studies have illustrated how more general situations can be formally specified and implemented. The result has always been a package or module that hides the details of the implementation and which can be used with confidence by an application programmer.

In all our examples we have discussed either well known data structures — **stacks, queues, trees,** and so on — or situations in which there was an easily identified data structure together with a set of equally easily identifiable operations. In reality, of course, the solution to a problem may involve several abstract data types, and these may not correspond very closely to the abstract data types discussed in this book. It turns out, however, that the very idea of an abstract data type is useful in program design, and leads to a program design technique called *object-oriented design.*

When faced with designing a program to solve a real problem the technique is to *start* by identifying the abstract data types which occur naturally in the problem. That is, the *operations* which define the abstract data types are identified first. In this approach it has become popular to refer to abstract data types as *objects,* and the program designer's task is that of identifying and defining objects. To define an object means to specify the operations in which that object takes part. Thus, from this point of view, an object *is* an abstract data type.

Here is an example which shows how this technique works. The problem is quite simple but the technique reveals aspects of the solution which may not be immediately apparent using other techniques of program design.

9.2. Example: maintaining a bank account

9.2.1 The problem

The program to be designed is to maintain a single bank account. The user of the program is to be able to ask, through a system of options, for a sequence of activities to be performed: display the current balance, deposit an amount of money into the account, make a debit to the account by some amount, add a month's interest to the account (at a rate of 1.5% per month), and stop the execution. To do this the program should prompt the user with a menu which provides a list of the available options:

B	display the current balance
P	make a payment (deposit) into the account
D	debit the account
I	add a month's interest to the account
Q	quit (halt processing)

All the necessary data, the amounts of deposits, and of debits, are to be input from a terminal by the user.

9.2.2 Identify the key objects

To identify the key objects we must answer the question: 'what things (objects) are mentioned in the problem description?' That is, we pick out the names of objects from the problem description — they are usually nouns — and build up an *object table* which, for the present, will contain just names and descriptions. The description of an object should be sufficient to identify the **type** of the object. For the bank account problem we get *Table 9.1*.

In this example, the notion of *bank account* incorporates all the other objects in the object table. A bank account is a kind of abstract data type that is defined in terms of a collection of other abstract data types. That is, once we have defined the remaining objects we shall, as a result, have specified the *bank account* object. Thus, *bank account* refers to the whole problem and, therefore, we shall henceforth omit it from the object table.

Table 9.1   An object table

Object	Description
Bank account	An amount of money maintained by a bank
Option	Characters input by the user to indicate an action
Current balance	The amount of money currently in the account
A deposit	An amount of money added to the account
A debit	An amount of money withdrawn from account
Monthly interest	An amount of money added to the account
A menu	A print out of available options

9.2.3   Identify the operations

To identify the operations we must answer the question: 'what activities (operations) can the objects take part in according to the problem description?' That is, we pick out the actions in the problem description — they are usually verbs — and add this information to the object table. At this point, to save space, it will be convenient to reduce the description part of the object table to show just the type of each object. The updated object table is shown in *Table 9.2*.

Table 9.2 shows the final situation after some considerable thought has taken place. It is not suggested that the operations listed there would be written down immediately. It is unlikely, for example, that the designer would recognize immediately that *calculate interest* would be an operation that the object *current balance* would take part in. However, when considering the object *monthly interest* it might become clear that the calculation of *monthly interest* involves *current balance*. It is at this stage that the operation *calculate interest* would be added to the operations listed against *current balance* (see [3] in *Table 9.2*). Hence, when adding an operation to the object table thought must be given to which other objects are involved in that operation. The design process is, therefore, one in which the lists of operations are developed simultaneously.

When developing an operations list it is important that the designer keep in mind any input or output operations that each object will take part in (see [2] in *Table 9.2*). Similarly, consideration must also be given to initialization operations. In our example, the objects *payment, debit,* and *option* are all initialized through input operations. The object *monthly interest* is calculated from the value of other objects and does not need to be otherwise initialized. The one exception is *current balance* because it is the only object in our example which is updated (i.e. changed by operations) and hence must be explicitly initialized (see [1] in *Table 9.2*).

Table 9.2   An extended object table

Object	Description	Operations
Option	Character	Read in from the terminal
Current balance	Amount of money	Add a deposit to current balance Add interest to current balance Take a debit from current balance Initialize current balance [1] Write out current balance [2] Calculate interest [3]
Deposit	Amount of money	Read in from the terminal Add to current balance
Debit	Amount of money	Take from current balance Read in from terminal
Interest	Amount of money	Calculate interest Add to current balance
Menu	Terminal listing	Write out to terminal

It should be quite clear by now that an individual operation might be listed against more than one object. That is, it is possible for several objects to be involved in a single operation.

For convenience, we shall now rewrite the object table in terms of the operations to show which objects are involved in each operation. It is also be useful to indicate whether the objects are changed by the operation, so we have added a further column for this purpose.

Table 9.3   An operations table

Operation	Object	Type	Changed?
Add a deposit	Current balance	Amount of money	yes
	Deposit	Amount of money	no
Add interest	Current balance	Amount of money	yes
	Interest	Amount of money	no
Take debit	Current balance	Amount of money	yes
	Debit	Amount of money	no
Initialize Current balance	Current balance	Amount of money	yes
Write out Current balance	Current balance	Amount of money	no
Calculate interest	Current balance	Amount of money	no
	Interest	Amount of money	yes
Read in deposit	Deposit	Amount of money	yes
Read in debit	Debit	Amount of money	yes
Read in option	Option	Character	yes
Write out menu	Menu	Listing	no

9.3   Implementation

The next step is implementation: the writing of one procedure (or function, as appropriate) for each operation. The important points to remember at this stage are:

(i)   choose a name for each procedure which reflects the operation that it implements;

(ii) the parameters of the procedures are objects;

(iii) the types of the parameters are the types of the objects.

At this stage the types of the objects should be as 'natural' as possible. No thought should have been given to devising suitable representations for the types. In this example we are primarily interested in manipulating amounts of money and we should not prejudge how such objects are to be stored (e.g. as real values, pairs of integers, strings, or whatever).

Hence, the headings for the operations are easily written down and incorporated into a definition module:

```
DEFINITION MODULE BankAccount;
    TYPE Money;
    PROCEDURE AddDeposit ( VAR CurrentBal: Money; Deposit: Money);
    PROCEDURE AddInterest ( VAR CurrentBal: Money);
    PROCEDURE TakeDebit ( VAR CurrentBal: Money; Debit: Money);
    PROCEDURE InitBalance ( VAR CurrentBal: Money);
    PROCEDURE WriteBalance ( CurrentBal: Money);
    PROCEDURE CalculateInterest ( VAR Interest: Money; CurrentBal: Money);
    PROCEDURE ReadDeposit ( VAR Deposit: Money);
    PROCEDURE ReadDebit ( VAR Debit: Money);
    PROCEDURE ReadOption ( VAR Option: CHAR );
    PROCEDURE WriteMenu;
END BankAccount.
```

The type *Money* is an opaque type and its definition will be resolved once we look at the implementation module. For the present, however, we shall examine the application program which can now be designed using these operations. Here, the techniques of top-down design are appropriate. Now, of course, we design by making as much use of the given operations as possible. Thus, one possible design is listed below.

```
initialize current balance
repeat
    write out menu
    read option
    case option of
        'B':  write out balance
        'D':  read in debit
              debit current balance
        'I':  add interest to current balance
        'P':  read in deposit
              add deposit to current balance
        'Q': quit
    end case
until option is quit
```

Such a design can be converted to a module in a very straightforward manner since each step can be converted to the call of a known procedure. For our current purpose we shall assume that the operation *readoption* will validate its input so that only valid input will be passed to the case statement. Hence we obtain the following main module:

```
MODULE Bank;

FROM BankAccount IMPORT Money, InitBalance, WriteMenu, ReadOption,
      WriteBalance, ReadDebit, AddInterest, ReadDeposit, AddDeposit;
FROM InOut IMPORT WriteString, WriteLn;

VAR
   Balance: Money;
   Option: CHAR;

BEGIN
   InitBalance (Balance);
   WriteMenu;
   REPEAT
      ReadOption (Option);
      CASE Option OF
         'B':  WriteBalance (Balance) |
         'D':  ReadDebit (Debit);
                  TakeDebit (Balance, Debit) |
         'I':     AddInterest (Balance) |
         'P':     ReadDeposit (Deposit) |
                  AddDeposit (Balance, Deposit) |
         'Q':  (* Quit *)
      END
   UNTIL Option = 'Q'
END Bank.
```

The application program has been written completely in terms of the operations of the abstract data type *BankAccount*.

It is now time to turn our attention to the implementation of the operations, and our first task would normally be to find a suitable representation for the type *Money*. However, for illustrative purposes, we shall suppose that some other programmer has had the foresight to provide an implementation of an abstract data type dealing with operations on money. This is not an unreasonable assumption since, in a typical computer environment, many applications will involve money calculations, and such an abstract data type would be likely to exist. Here is the definition module for such an abstract data type.

DEFINITION MODULE *MoneyMatters;*
TYPE *Money;*

PROCEDURE *InputMoney (***VAR** *Amount: Money);*

PROCEDURE *OutputMoney (Amount: Money);*

PROCEDURE *AddMoney (Amount1, Amount2: Money): Money;*
(* *result is Amount1 plus Amount2* *)

PROCEDURE *SubtractMoney (Amount1, Amount2: Money): Money;*
(* *result is the difference Amount1 minus Amount2* *)

PROCEDURE *MultMoney (Amount: Money; Factor: Real): Money;*
(* *multiply Amount by the scalar Factor* *)

PROCEDURE *InitializeMoney (): Money;*
(* *The result is a zero amount of money* *)

PROCEDURE *EqualMoney (Amount1, Amount2: Money): Booolean;*
(* *If Amount1 = Amount2 then true else false* *)

PROCEDURE *LessMoney (Amount1, Amount2: Money): Boolean;*
(* *If Amount1 < Amount2 then true else false* *)

PROCEDURE *GreaterMoney (Amount1, Amount2: Money): Boolean;*
(* *If Amount1 > Amount2 then true else false* *)

END *Money Matters.*

The implementation module for *BankAccount* can now be written:

IMPLEMENTATION MODULE *BankAccount;*
FROM *MoneyMatters* **IMPORT** *AddMoney, SubtractMoney,*
                                InitializeMoney, MultMoney;
FROM *InOut* **IMPORT** *WriteString, WriteLn;*

PROCEDURE *AddDeposit (***VAR** *CurrentBal: Money; Deposit: Money);*
BEGIN
    CurrentBal : = AddMoney (CurrentBal, Deposit)
END *AddDeposit;*

PROCEDURE *AddInterest (***VAR** *CurrentBal: Money);*
BEGIN
    Interest : = MultMoney (CurrentBal, 1.5);
    AddMoney (CurrentBal, Interest)
END *AddInterest;*

```
PROCEDURE  TakeDebit (VAR  CurrentBal: Money;  Debit: Money);
VAR
   Temporary: Money;
BEGIN
   Temporary : = Subtract (CurrentBal, Debit);
   IF Temporary < InitializeMoney
   THEN
      WriteString ('Insufficient funds to perform debit.');
      WriteLn
   ELSE
      CurrentBal : = Temporary
   END
END TakeDebit;


PROCEDURE  InitBalance ( VAR  CurrentBal: Money);
BEGIN
   CurrentBal : = InitializeMoney;
END InitBalance;


PROCEDURE  WriteBalance (CurrentBal: Money);
BEGIN
   WriteString ('The Current balance is ');
   OutputMoney (CurrentBal);
   WriteLn;
END WriteBalance;


PROCEDURE  CalculateInterest (VAR Interest: Money;
                                   CurrentBal: Money);
BEGIN
   Interest : = MultMoney( CurrentBal, 1.5);
END CalculateInterest;


PROCEDURE  ReadDeposit ( VAR Deposit: Money);
BEGIN
   WriteString ('Enter the deposit amount: ');
   InoutMoney (Deposit);
END ReadDeposit;


PROCEDURE  ReadDebit ( VAR Debit: Money);
BEGIN
   WriteString ('Enter the debit amount');
   InputMoney ( Deposit);
END ReadDebit;


PROCEDURE  ReadOption ( VAR Option: CHAR);
BEGIN
   WriteString ('Enter option character: ');
   Read (Option);
END ReadOption;
```

```
PROCEDURE WriteMenu ();
BEGIN
    WriteLn;
    WriteString ('The Bank Account Program');
    WriteLn;
    WriteString ('The available optionms are:');
    WriteString ('B: to write out current balance');
    WriteString ('D: to debit the account');
    WriteString ('I: to add to a month's interest to the account');
    WriteString ('P: to add to a deposit to the account');
    WriteString ('Q: to quit processing.');
    WriteLn;
END WriteMenu;
END BankAccount.
```

9.4 Review

It should be clear from the above that, because of the concentration on operations, object-oriented programming leads to highly modular programs. This is particularly useful when dealing with large programs. The technique also encourages the applications programmer to make as much use of pre-existing modules as possible, with obvious economic advantages.

The major alternative to object-oriented programming is top down design. Top down design is more concerned with program control structures and is therefore more suited to evolving the implementation of complex operations, with the result that it does not give much help in deciding how to modularize a design.

The two techniques, object-oriented programming and top-down design, are probably best used in tandem, where the programmer uses each technique for a different purpose: (i) by using object-oriented programming to identify objects and their operations to develop the modular structure of the program, and (ii) by using top-down design to evolve the more complex operations (as well as the main program). An advantage of this approach is that is tends to lead to main programs which are very readable and easy to understand because they consist mainly of procedure calls which closely resemble the abstract operations of the problem being solved.

An important advantage of object programming is that it leads to more easily maintainable programs. The main reasons for this are:

(i)   the procedures are usually short and readable because they perform single, clearly defined tasks;

(ii) it is easier to locate and modify an operation when it is confined to a single procedure.

Exercise 9.1

From the following informal description of a computer system, pick out the objects and the operations and build up an operations table similar to *Table 9.3*.

A mail order company keeps all its stock in a single warehouse. The stock consists of a large number of different items, and the range of items varies as new items are added to stock and old items are no longer replenished. Items are added to stock when deliveries are made. Each delivery is accompanied by a delivery note specifying the quantity of each item in the delivery. Goods are removed from the warehouse by filling orders. An order specifies the quantity of each item required. For each item the warehouse keeps a price and a re-order level (when stocks fall below this level additional items have to be delivered to the warehouse). The warehouse manager needs a program that will:
  (a)   maintain the current stock level;
  (b)   print out the current stock level of each item;
  (c)   issue requests for more stock when the re-order level is reached;
  (d)   print out the total value of stock.

9.5   Object-oriented programming languages

Interest in the use of objects in programming has increased so dramatically that a large number of programming languages have been designed and implemented around this concept. The majority of these *object-oriented programming languages* have their roots in one pioneering language named *Smalltalk,* developed at the Xerox Palo Alto Research Centre, USA. Smalltalk was originally seen as a language for developing user interfaces on personal computers and has led to the ideas behind the notion of the *desk top,* a user interface based on the idea of several documents lying on a desk, the documents being represented by a series of *windows* drawn on the screens of personal computers. Such a system has been implemented on the Apple Macintosh series of personal computers. It was felt that the majority of programming languages, having been developed for professional programmers working in specialized areas, were not suited for the typical personal computer user. The conclusion was that a radically different approach was needed.

The new approach sees programming as instructing 'real' objects to perform a sequence of actions. For example, to obtain diagrams on a computer screen is achieved by instructing a 'computerized' pen to draw the figures. Such an approach views a pen as an *object* which is capable of performing certain tasks (or, alternatively, a pen 'knows' how to carry out a set of operations). Typically a pen can draw lines, move to specified places on a screen, traverse the screen in different directions, and so on. Thus, a pen is an object which can be recognized as such by the operations it takes part in: in other words it is an abstract data type. But, as in real life, there exist many pens, some of which are identical, and some have very different characteristics. Thus, *pen* is a generic concept. This leads us to define an object to be a specific instance of a generic description (in the same way that one can obtain instances of generic packages in Ada). In an object oriented programming language, therefore, the programmer would be able to say such things as, create a new instance of *pen* which has the following characteristics: it uses blue ink and draws thin lines'. Thereafter, the programmer would be able to use this particular pen to draw a particular diagram. Of course, for other diagrams totally different pens might be required.

In more concrete computing terms, an object is a collection of private data and a set of operations (procedures). The private data can be accessed only through these operations — which is why the data is private. This idea is, of course, central to information hiding. In the case of a particular pen, the private data might include information about the colour of its ink, or its current position on the screen. The only way to access or modify such data is through actions that the pen knows about. For example, the current position will be updated when the pen is told, via one of its procedures, to move to a new position on the screen.

We have been using rather unusual terminology to describe objects. For example, we have used phrases like: 'an object *knows* how to do things', and 'telling an object to do something'. The reason is simple.

Object oriented programming focuses primarily on the objects themselves rather than on the mechanics of how they achieve their goals. To emphasize this approach, the instructions given to objects are in the form of *messages* which inform the object which of its associated actions is to be carried out.   The model behind object-oriented programming is one in which objects interact by passing messages. Here is an example of this approach.

Suppose you want to draw a square at a particular position on the screen of your computer. The first thing you would do is to obtain a pen to draw with, and a typical statement to do this is:

> *yourpen := pen new: thin*

This statement causes a *new* object, called *yourpen,* of type *pen,* and which draws *thin* lines, to be created. The identifier *yourpen* is used to identify this particular pen in the rest of the program. The identifier *pen* is the name of a generic object that understands that the message *new: thin* means create a specific instance of a pen which draws thin lines. Here we are not using any specific object-oriented language but our syntax is typical of the *genre.*   First the object of interest (*pen*) is mentioned, followed by a message (*new: thin*). A message contains the name of an action (*new*) followed by arguments (*thin*), if any. Having obtained a pen (*yourpen*), you can now instruct it to carry out its drawing tasks. To draw a square of side 200 units and whose top left hand corner is at position (100, 200) on the screen (see *Fig. 9.1*), the following sequence of messages could be used:

yourpen colour: blue	*{fill pen with blue ink}*
yourpen moveto: 100, 200	*{move pen to a specific position without drawing}*
yourpen drawlineto: 300, 200	*{draw a line from current position to the new position (300, 200)}*
yourpen drawlineto: 300, 400	
yourpen drawlineto: 100, 400	
yourpen drawlineto: 100, 200	
yourpen moveto: 0, 0	*{move pen out of way}*

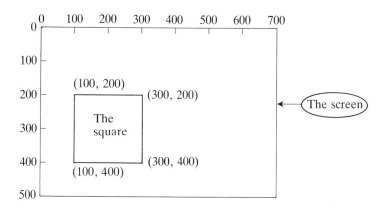

Fig. 9.1 The screen: showing the coordinates of the square

In this example it is clear that a specific pen must know about the operations *colour, moveto,* and *drawlineto,* and these would be part of the definition of *pen.*

Such sequences of messages form a subprogram; it is possible to generalize them because the square that has been drawn is obviously just one of many suares that could be drawn on a screen (at different locations and with different sizes). A question that naturally arises is whether it would be possible to generalize the concept of *square* and hence define a new object.

To define *square* we must first identify the operations that such an object can take part in. Here we shall define three operations, given informally below:

draw: draws a new square at location (x, y), the coordinates of its top left hand corner, with size of side *size,* and using a specific pen *apen;*

erase: removes the square from the screen;

stretch: makes a current square larger or smaller by giving it a new size of side. The new size is calculated from the old size by adding a new *amount* to it.

There is one trick associated with drawing on a screen that makes the definition of these operations quite easy. We shall assume that, before any drawing is done, the screen is 'blank'. That is, the screen is of a uniform colour with no markings on it. The colour of a blank screen is stored in a variable named *background*. Hence, if any object is drawn on the screen using ink with colour *background* then that object will not be visible. This means that if we draw on top of an existing square using background colour the effect will be to erase that square from the screen. *Fig. 9.2* shows how the three *square* operations might be defined in a typical object oriented language. In object-oriented language terminology, a generic abstract data type is called a *class*.

```
class square;
instance variables: x, y, size, apen;
instance methods:
  draw: ink
     apen colour: ink
     apen moveto: x, y
     apen drawlineto: x + size, y
     apen drawlineto: x + size, y + size
     apen drawlineto: x, y + size
     apen drawlineto: x, y
     apen moveto: 0, 0;

  erase
     draw: background;

  stretch: amount
     self erase
     size := size + amount
     self draw: ink;

end class
```

Fig. 9.2 A *class* definition

There are several new ideas in this *class* definition.

(i)   Whenever a new instance of *square* is created it must have a physical location on the screen. This information is held in its *instance variables*. Instance variables are equivalent to local variables and each instance of *square* will have its own private set of instance variables.

(ii)   The *instance methods* are those procedures that describe the behaviour of the instances when they receive messages. It is only necessary to have one set of such procedures since they can be shared between all instances.

(iii) *ink* is an argument to *draw*. Therefore, we have defined *erase* to be the same as *draw* but using background ink.

(iv) The object *self* is used to refer to the particular instance itself. In general there can be several squares in existence at any time, and it is important to be able to refer to them individually. Normally, the programmer would invent separate identifiers for each instance. However, when defining a class, we would not know what names a programmer is subsequently going to use. Therefore, a way has to be found of indicating that, whatever the name of the instance might be, it is that specific instance that is being referred to.

(v) The procedures *draw* and *stretch* have a single argument each named *ink* and *amount,* respectively.

There is one procedure missing in this definition: the procedure that creates new instances. Whenever the message *new* is sent to the class *square* a new, initialized instance must result. For example, we might like to create a new square at location (250, 375), with side size 20, and using *yourpen*. This can be achieved with the statement:

```
square  new: 250, 375, 20, yourpen
```

This is equivalent, in other procedural languages, to calling a procedure and specifying its actual parameters. Hence, a procedure named *new* must be included in the definition of *square*.

These operations can be used in applications by such statements as:

```
asquare := square new: 100, 200, 200, yourpen
asquare draw: blue
asquare stretch: 50
asquare erase
```

Here, a new square (*asquare*) is created (its instance variables are initialized), the new square is drawn on the screen in blue ink, it is then stretched by 50 units, and finally erased.

Hence, at the abstract level an object is equivalent to an abstract data type. At the concrete level, in programming languages, there is a difference in emphasis. An abstract data type is implemented as a set of procedures operating on private data, and a programmer uses the procedures to implement an application. Here the programmer views a program as a sequence of calculations. Conversely, an object-oriented programming language provides the ability to define objects, and a programmer sends messages to them viewing the processing as one in which objects carry out tasks. However, both schemes require the same information to be passed between the application and the implementation.

Exercise 9.2

In a procedural language like Ada, at what stage, i.e. during compilation or run-time, does an instance of an abstract data type come into existence, and when are its private data initialized? At what stage(s) do the equivalent actions take place for an object oriented language?

Although we have not explicitly shown it, object-oriented programming languages do provide a high level of information hiding. Their syntax is such that users are allowed to know about the messages that can be sent to an object, and about what additional information may be required in so doing. There is, however, no visibility of private data or of the implementations of the operations.

Object-oriented programming languages have advanced the cause of re-usability through the concept of *inheritance*. Inheritance is *not* an essential feature of *objects* but it is of considerable value. It is the object-oriented approach that enables easy implementation of the idea.

The fundamental idea of inheritance is simple: if you want to define a new object, and you know that it is similar to an existing object, then define it in terms of the existing object. For example, suppose that we already have an object called *rectangle* for which the operations of *draw, erase,* and *stretch* have been defined. It would be simpler to define *square* as a *rectangle* whose sides were of the same length. Typically, an object-oriented programming language will provide facilities to define the class *square* by refering to the class *rectangle*. In so doing we can say that the new class, *square,* should have the same operations as *rectangle*. That is, *square inherits* the operations of *draw, erase,* and *stretch*. This means that we do not have to rewrite these procedures, we are just reusing existing software.

Check point

(i)   At the abstract level, what is an *object?*

(ii) In an object-oriented programming language, what is an *object?*

(iii) What, in an object-oriented language, is equivalent to the following in a conventional procedural language like Ada:
   (a)   calling a procedure;
   (b)   actual parameters;
   (c)   formal parameters;
   (d)   a generic package?

(iv) What is inheritance?

Solutions

(i)   An abstract data type. An object is defined by a set of operations.

(ii) A set of operations acting on a collection of private data. An object is something that knows about, and is capable of taking part in, a set of actions.

(iii) (a)   sending a message;
   (b)   arguments in a message;
   (c)   formal arguments in the definition of an action;
   (d)   a *class*.

(iv) Inheritance means using already existing software. Inheritance is a way of defining new objects from existing ones.

References and bibliography

Aho, V. A., Hopcroft, J. E. and Ullman, J. D. (1983). *Data structures and algorithms*. Addison-Wesley, London.

Backhouse, R. C. (1986). *Program construction and verification*. Prentice-Hall, Englewood Cliffs, New Jersey. A mathematical text on specification and program proving.

Bishop, J. (1986). *Data abstraction in programming languages*. Addison-Wesley, Wokingham. A broadly based discussion of problems associated with abstraction in programming languages.

Caverly, P. and Goldstein, P. (1986). *Introduction to Ada*. Brooks/Cole, Monterey, California.

Clark, R. and Koehler, S. (1982). *The UCSD Pascal handbook*. Prentice-Hall, Englewood Cliffs, New Jersey.

Cleveland, J. C. (1986). *An introduction to data types*. Addison-Wesley, Wokingham. A programming language based discussion of types, type checking, abstract data types, polymorphism, and specifications.

Cox, B. J. (1986). *Object oriented programming*. Addison-Wesley, Wokingham. Shows how to build a library of reusable objects and constructs several sample applications. Based on an object-oriented version of the C language.

Date, C. J. (1986). *An Introduction to Database Systems,* (4th edn). Addison-Wesley, Wokingham. One of *the* authoritative texts on database systems. Particularly strong in its treatment of the relational data model.

Gehani, N. (1986). Specifications: formal and informal — a case study. In *Software specification techniques* (ed. N. Gehani and A. D. McGettrick). pp.173–185. Addison-Wesley, Wokingham. Compares the informal specification of a real system with its algebraic specification. Reaches the conclusion that 'formal specifications cannot replace informal specifications — they are complementary ... informal specifications are easier to read and understand while the formal specifications tend to be clearer, precise, unambiguous, etc.'

Gehani, N. and McGettrick, A.D. (eds) (1986). *Software specification techniques*. Addison-Wesley, Wokingham. A collection of recent papers on specification. Includes accounts of a variety of approaches, such as Z, VDM, OBJ and GYPSY.

Guttag, J. V. (1977). Abstract data types and the development of data structures. *Communications of the ACM.* 20:307–315. A very accessible paper by one of the pioneers in the area. Guttag's Ph.D. thesis was entitled 'The specification and application to programming of abstract data types'.

Guttag, J. V., Horowitz, E. and Musser, D. R. (1978). The design of data type specifications. In *Current trends in programming methodology. Volume IV. Data structuring* (ed. R. T. Yeh). pp.60–79. Prentice-Hall, Englewood Cliffs, New Jersey. An excellent treatment of the axiomatic approach. Other chapters in the volume are also of interest, in particular that of T. A. Standish and that of J. A. Goguen, J. W. Thatcher, and E. G. Wagner.

Horowitz, E. and Sahni, S. (1976). *Fundamentals of data structures*. Pitman, London. A compendious textbook that includes many examples of axiomatic specifications. Marred by the choice of a somewhat arcane algorithm description language in discussing implementations.

Jones, C. B. (1980). *Software development: a rigorous approach*. Prentice-Hall, London. A seminal text in the appreciation of the benefits of formal methods for software construction.

Jones, C. B. (1986). *Systematic software development using VDM*. Prentice-Hall, London. An advanced mathematical text on specifications, program proving and data types. Can be seen as a re-working of Jones (1980) in the light of current developments and experience.

Liskov, B. H. and Zilles, S. N. (1975). Specification techniques for data abstractions. *IEEE Transactions on Software Engineering*. 1:294–306.

MacLennan, B. J. (1983). *Principles of programming languages*. Holt, Rinehart, and Winston, New York. Discusses the design of several languages, evaluates the languages, and looks at details of their implementation. Has a chapter on object-oriented programming based on Smalltalk. Gives a historical perspective.

Martin, J. M. (1986). *Data types and data structures*. Prentice-Hall, London. Data structures presented as implementation structures of abstract data types. Detailed and rigorous discussion of algebraic specification, implementation, and verification.

Schmucker, K. J. (1986). *Object-oriented programming for the Macintosh*. Hayden, Hasbrouck Heights, New Jersey. Shows how Pascal can be developed into an object-oriented language.

Sommerville, I. (1985). *Software engineering*. (2nd edn). Addison-Wesley, Wokingham.

Stone, R. G. and Cooke, D. J. (1987). *Program construction*. Cambridge University Press, Cambridge. Considers the disciplined construction of procedural programs from formal specifications. Deals with program construction, abstract data types, program design language, and verification.

Stroustrup, B. (1987). What is object-oriented programming? *Proceedings of the European conference on object-oriented programming*. Paris. pp. 57-76.

Stubbs, D. F. and Webre, N. W. (1985). *Data structures with abstract data types and Pascal.* Brooks/Cole, Monterey, California. A first course in data structures. Semi-formal approach to specification, using structured English to describe pre- and post-conditions for each operation defining the abstract data type. Syntax specified by Pascal procedure headings. Program construction using modules and packages. Implementation and (particularly) performance are considered. Pascal is a prerequisite. Starts with arrays, records, and pointers and goes on to lists, stacks, queues, trees, sets, strings, and graphs. Sorting and hashing algorithms are included and there are case studies. Final chapter looks at language facilities (e.g. generic packages, separate compilation, modules). Appendix gives formal specification using predicate calculus of stack, queue, priority queue, set, and list. A student-oriented book; good layout, useful marginal notes and diagrams.

Texel, P. P. (1986). *Introductory Ada.* Wadsworth, Belmont, California. An excellent introduction to Ada. Begins with the package concept and introduces language features at a later stage. Does not cover the entire language but is sufficient for a first course.

Walker, B. K. (1986). *Modula-2.* Wadsworth, Belmont, California. Useful informal descriptions of MODULA-2 concepts.

Walker, R. and Adelman, C. (1976). Strawberries. In *Explorations in classroom observation* (ed. M. Stubbs and S. Delamont). pp.133–150. John Wiley, London.

Wirth, N. (1985). *Programming in MODULA-2.* (3rd edn). Springer-Verlag, Berlin. The current language definition.

Wittgenstein, L. (1958). *Philosophical investigations.* Basil Blackwell, Oxford.

Zimmer, J. A. (1985). *Abstraction for programmers.* McGraw-Hill, New York. Concentrates on programming concepts rather than data abstraction.

Solutions to exercises

Chapter 2

2.1 A possible function body is

if $y >= x$
then
   Maximum := y
else
   Maximum := x

where the order of comparison between the two source data items in the original version is reversed.

2.2 The syntax of **pop** is that its source data is a stack and its result is a stack. Notice that although **pop** deletes an item, that item is not returned as a result and therefore does not appear in the syntax of **pop**.
   The syntax of **push** is that its source data is an item and a stack, and its result is a stack.

2.3 The semantics of **pop** are that it deletes the top item from the source data stack and returns the changed stack as the result.
   The semantics of **push** are that it inserts the source data item on the top of the source data stack to yield the stack required as the result.

2.4 **replicatestack:** $S \rightarrow S$

2.5 **push:** $I \times S \rightarrow S$

The source data component is defined as a member of the Cartesian product of *I* and *S:* that is, a pair consisting of an item and a stack, in that order.

2.6 **mergestacks:** $S \times S \rightarrow S$

2.7 **subtract:** $Z \times Z \rightarrow Z$

Notice that the 'meaning' of **subtract** is not obvious from the syntax and, also, that in the source data set (the pair of integers) it is not possible to indicate that the second integer is **subtract** ed from the first. The ordering of this pair is, of course, crucial, but its exact significance would only be apparent from the semantics of **subtract**.

2.8 **isemptystack:** S → B

2.9 **top**, as before, produces as its result **one** element from the union of *I* and *M* (which *particular* element is defined, later, by the semantics of **top**). In contrast, **anothertop** produces as its result a pair, consisting of an item **and** a message, in that order.

2.10 The set *M* would consist of the members *the employee stack is empty* and *the employee stack is full* (or a similar string). The syntax of **pushemployeestack** would become

    pushemployeestack: Y × P → P ∪ M

2.11 **pop (createstack)** = **createstack**   *(S5)*

2.12 **replicatestack** (s) = s

2.13 **top (createstack)** = *the stack is empty*   *(S6)*

2.14 **pop (createstack)** = *the stack is empty*   *(S5')*

2.15 A stack which can be pictured as:

2.16 **push**(a, (**push**(c, **createstack**)))

2.17 *Axiom L7* can be regarded as applying **first** to a list which cannot be empty. That is, the source data for **first** is **concatenate** (**make**(i), a)) which will never result in an empty list, irrespective of the value of *a*. **first** applied to this list is then defined to be *i*. *Axiom L8* defines the application of **first** to an empty list by using as the source data a list which is always empty: that produced by **createlist**.

2.18 First, the application of length is written in the style of the left-hand side of *axiom L12* to give:

    length (concatenate (make(Richard), <Madie, Edna>))

which *axiom L12* states is equal to

1 + **length** (<Madie, Edna>)

which, following the main text, will evaluate to *3,* that is 1 + 1 + 1 + 0.

2.19 (i) *i*   (ii) **createlist**   (iii) **createlist**   (iv) *i*   (v) *false*

2.20 pre-**minimum** (a, b) ::= *true*
post-**minimum** (a, b; r) ::= (r <= a) ∧ (r <= b) ∧ (r = a ∨ r = b)

2.21 pre-**subtract** (x,y) ::= *true*
post-**subtract** (x,y; r) ::= x = y + r

2.22 pre-**subtract** (x,y) ::= (x > 0) ∧ (y > 0) ∧ (x ≥ y)

2.23 pre-**top** (s) ::= (s ≠ **createlist**)
post-**top** (s; r) ::= r = **first** (s)

An alternative pre-condition is:
pre-**top**(s) ::= **not** (**isemptylist** (s)).

The syntax is now
top: S → I

2.24 The syntax and semantics of **push** are affected, since it is the only operation which can increase the size of a stack. The syntax of **push** becomes

push: I × S → S ∪ M

and the set *M* now includes the member *the stack is full.* The pre-condition for **push** remains the same, but the post-condition becomes

post-**push**(i, s; r) ::=   *if* **length** (s) = 10
                                 then
                                         r = *the stack is full*
                                 else
                                         r = **concatenate** (**make** (i), s)

An alternative for the right-hand side of the post-condition is

if **length** (s) < 10
then
    r = **concatenate** (**make** (i), s)
else
    r = *the stack is full*

The required change cannot be handled by amending the pre-condition (so that, for example, the source data stack must not contain more than 9 items) since we need to be able to able to produce the message *the stack is full* when the constraint is exceeded. A somewhat more elegant solution is obtained by adding to the *SETS* entry the following

> C   set of integer constants consisting of the single member *maxitems* whose value is *10*.

The post-condition then becomes

post-**push**(i, s; r) ::=   *if* **length** (s) = *maxitems*
                            then
                                 r = *the stack is full*
                            else
                                 r = **concatenate** (**make** (i), s).

2.25 *NAME*
   otherstack (item)

SETS
   O   set of otherstacks
   I   set of items
   B   set consisting of the Boolean values *true* and *false*
   M   set of message values consisting of the single member
       the otherstack is empty

SYNTAX
   createotherstack: → O
   top: O → I ∪ M
   pop: O → O ∪ M
   push: I × O → O
   isemptyotherstack: O → B
   bottom: O → I ∪ M

SEMANTICS
   pre-**createotherstack** ( ) ::= *true*
   post-**createotherstack** (o) ::= o = **createlist**
   pre-**top** (o) ::= *true*
   post-**top** (o; r) ::= *if* o = **createlist**
                   then r = *the otherstack is empty*
                   else r = **first** (o)
   pre-**pop** (o) ::= *true*
   post-**pop**(o; r) ::= *if* o = **createlist**
                   then r = *the otherstack is empty*
                   else r = **trailer** (o)

pre-**push** (i, o) ::= *true*
post-**push** (i, o; r) ::= r = **concatenate** (**make** (i), o)
pre-**isemptyotherstack** (o) ::= *true*
post-**isemptyotherstack** (o; b) ::= b = **isemptylist** (o)
pre-**bottom** (o) ::= *true*
post-**bottom** (o; r) ::= *if* o = **createlist**
　　　　　　　　　　　then r = *the otherstack is empty*
　　　　　　　　　　　else r = **last** (n)

invariant assertion
　*Whenever an operation is applied to a value from M then the
　result of the operation is that same value from M.*

An alternative for o = **createlist** is **isemptylist** (o).

Chapter 3

3.1 A specification of an abstract data type is a definition which is
independent of representation or implementation details. A specification
describes the operations which define the abstract data type independently of
how those operations are carried out. A respresentation describes the data
structures which will be used in the subsequent implementation.   A
representation is independent of implementation details of a particular
programming language. An implementation is the program code which
carries out the operations.

3.2　(i)　**var** *NameStack: Stack;*
　　　(ii)　*NameStack.Item* [*NameStack.Top – 1*]
　　　(iii)　*NameStack.Top = 0.*

3.3 The messages in *Top* and *Pop* arise from inherent constraints; the one
in *Push* arises from a representation constraint.

3.4　**procedure** *CreateStack (* **var** *S: Stack);*
　　　begin
　　　　S := **nil**
　　　end *{CreateStack};*

3.5　**function** *IsEmptyStack (S: Stack): Boolean;*
　　　begin
　　　　IsEmptyStack := (S = **nil** *)*
　　　end *{IsEmptyStack};*

3.6 procedure *Pop (var S: Stack);*
    var
        Temp: Link;
    begin
        if *IsEmptyStack (S)*
        then
            Writeln (' The stack is empty.')
        else
            begin
                Temp := S;
                S := S ↑.Previous;
                dispose (Temp)
            end
    end *{Pop};*

3.7 No. The only two messages result from inherent situations.

3.8 Information will needed about how to invoke the five procedures. Their headings will suffice

    procedure *CreateStack (var S: Stack);*
    procedure *IsEmptyStack (S: Stack): Boolean;*
    procedure *Top (S: Stack; var Item: ItemType);*
    procedure *Pop (var S: Stack);*
    procedure *Push (Item: ItemType; var S: Stack);*

always provided, of course, that the purpose of each of the routines and the definitions of the parameters are known. As far as the application programmer is concerned, the type *Stack* enables stack objects to be declared in the application program, and the type *ItemType* is to be chosen to suit the application. Apart from some knowledge about how error situations are catered for, which we have conveniently ignored for the present, the applications programmer requires no other information.

3.9 (i) A procedure or function separates the syntax and semantics of the operation it implements. The syntax is embodied in the routine heading; the semantics are contained within the body of the routine.

(ii) An application programmer needs to be aware only of the routine headings and will not be influenced by the implementation of the operation.

(iii) It is not necessary to change the application if a change is made to the implementation, because the routine headings will remain the same. The complete program (application and abstract data type) will need recompiling, however.

(iv) The implementation will closely resemble the formal specification, making it easier to construct and maintain.

3.10 Information hiding
(i)   improves program development and maintenance by separating the application program from the implementation, and
(ii)   makes software more easily re-usable.

3.11 None! The calls to the stack routines would remain the same; the declaration of the variables *NumberStack* and *CurrentNumber* would remain the same. Of course, if the pointer-based implementation were held in a differently named unit from the cursor-based implementation, the uses statement would need to reflect this fact.

3.12 Advantages:
(i)   Separation of the implementation of an abstract data type from its use in an application program.

(ii)   Information needed by the application program is readily available in the interface part of the unit, and is separate from some of the other details held in the implementation part of the unit.

(iii)   Some details of the implementation can be placed in the implementation part of a unit and, as a result, cannot be accessed by a program that uses the unit.

(iv)   The unit can be compiled separately from the application program, and the implementation part can be changed without affecting the application program.

Disadvantages:
(i)   The interface and implementation parts of a unit cannot be physically separated and, therefore, the implementation details are always visible.

(ii)   Some implementation details, which should be hidden, have to be placed in the interface part and can, therefore, be used by the application programmer.

(iii)   The definition of the type of the items in the abstract data type is held within the interface part of the unit and becomes, therefore, part of the implementation and cannot be changed in the application program. Separate units are required for each instance of an abstract data type attempting to manipulate different types of item.

3.13 procedure *Push (Item: ItemType;* var *S: Stack;*
                               var *Status: StackStatusType);*
begin
if *S.Top = MaxSize*
then
  begin
    Writeln (' The stack is full.');
    Status := StackFull
  end
else
  begin
    S.Top := S.Top + 1;
    S.Item [S.Top] := Item;
    Status := Ok
  end
end *{ Push};*

3.14 The status values must be known to the application programmer so that the application program can be designed to take action appropriate to the status values returned by the abstract data type routines. Therefore, the ideal place to define them is in the interface part of the unit.

3.15 Not if a status value were to be returned. In these circumstances the routine would return both a status value and the value at the top of the stack. Functions are used to return one value only.

Chapter 4

4.1   package *EmployeeStack2* is
          subtype *ItemType* is *String(1 . . 5);*
          type *StackRecord;*
          type *Link* is access *StackRecord;*
            type *StackRecord* is
              record
                Item: ItemType;
                Previous: Link;
              end record;

          type *Stack* is *Link;*

          procedure *CreateStack (S:* out *Stack);*
          function *IsEmptyStack (S:* in *Stack)* return *Boolean;*
          function *Top (S:* in *Stack)* return *ItemType;*
          procedure *Pop (S:* in out *Stack);*
          procedure *Push (Item:* in *ItemType; S:* in out *Stack);*
          end *EmployeeStack2;*

4.2   **package** *EmployeeStack2* **is**
    subtype *ItemType* **is** *String(1 . . 5);*
    type *Stack* **is private** ;

    procedure *CreateStack (S:* **out** Stack);
    function *IsEmptyStack (S:* **in** *Stack)* **return** *Boolean;*
    function *Top (S:* **in** *Stack)* **return** *ItemType;*
    procedure *Pop (S:* **in out** *Stack);*
    procedure *Push (Item:* **in** *ItemType; S:* **in out** *Stack);*

  private
    type *Link* **is access** *StackRecord;*
    type *StackRecord* **is**
        record
         Item: ItemType;
         Previous: Link;
        end record;
    type *Stack* **is**
        record
         Top: Link;
        end record;
  end *EmployeeStack2;*

4.3   The specification part of the package becomes:

  generic
    type *ItemType* **is private**;

  package StackPackage2 **is**
    type Stack **is private** ;

    procedure *CreateStack (S:* **out** *Stack);*
    function *IsEmptyStack (S:* **in** *Stack)* **return** *Boolean;*
    function *Top (S:* **in** *Stack)* **return** *ItemType;*
    procedure *Pop (S:* **in out** *Stack);*
    procedure *Push (Item:* **in** *ItemType; S:* **in out** *Stack);*

  private
    type *Link* **is access** *StackRecord;*
    type *StackRecord* **is**
        record
         Item: ItemType;
         Previous: Link;
        end record;
    type *Stack* **is** *Link;*
  end *StackPackage2;*

4.4 Since *EmployeeStack2* uses access types (pointers) there is only one exception — attempting to access items in an empty stack. We shall call this exception *StackEmpty*. The following actions are needed.

(i)   In the package specification the declaration: *StackEmpty:* **exception;** must be added;

(ii)   In the package body the exception must be raised in the procedures *Top* and *Pop* using such statements as the one below.

```
if IsEmptyStack
then
    raise StackEmpty;
else
    -- whatever is appropriate
end if;
```

Chapter 5

5.1 *NAME*
    xqueue (item)

SETS
    X   set of xqueues
    I   set of items
    B   set consisting of the Boolean values *true* and *false*
    M   set of message values consisting of the four members *the xqueue is empty, the xqueue is too full, the xqueue is getting full, the xqueue is not getting full.*
    C   set of integer constants consisting of the two members *maxitems*, denoting the value *100,* and *warningitems*, denoting the value *75.*

SYNTAX
    createxqueue: $\rightarrow X$
    front: $X \rightarrow I \cup M$
    addtoxqueue: $I \times X \rightarrow X \cup M$
    deletefromxqueue: $X \rightarrow X \cup M$
    isemptyxqueue: $X \rightarrow B$
    sizewarning: $X \rightarrow M$

SEMANTICS
    pre-**createxqueue** ( ) ::= *true*
    post-**createxqueue** (x) ::= x = **createlist**

    pre-**front** (x) ::= *true*
    post-**front** (x; r) ::= *if* x = **createlist**
                             then r = *the xqueue is empty*
                             else r = **first** (x)

pre-**addtoxqueue** (i, x) ::= *true*
post-**addtoxqueue** (i, x; r) ::= *if* **length** (x) = *maxitems*
                                         then r = *the xqueue is too full*
                                         else r = (**concatenate** (x, **make** (i))
pre-**deletefromxqueue** (x) ::= *true*
post-**deletefromxqueue** (x; r) ::= *if* x = **createlist**
                                         then r = *the xqueue is empty*
                                         else r = **trailer** (x)
pre-**isemptyxqueue** (x) ::= *true*
post-**isemptyxqueue** (x; b) ::= b = **isemptylist** (x)
pre-**sizewarning** (x) ::= x ≠ **createlist**
post-**sizewarning** (x; r) ::= *if* **length** (x) < *warningitems*
                                         then r = *the stack is not getting full*
                                         else r = *the stack is getting full*

invariant assertion
Whenever an operation is applied to a value from M then the
result of the operation is that same value from M.

5.2   **front** applied to the queue depicted in *Fig. 5.4* may be written in the
form of the left-hand side of *axiom Q4* as

front (**addtoqueue** (4, x))   *(1)*

where x denotes the queue < 7, 1>. Using the right-hand side of axiom *Q4*,
Expression (1) is the same as

if **isemptyqueue** (x)
then
    4
else
    front (x)   *(2)*

That is, the required result is obtained by

front (x).   *(3)*

Using a similar strategy, *Expression (3)* may be written in the form of the
left-hand side of *axiom Q4* as

front (**addtoqueue** (1, y))   *(4)*

where y denotes the queue < 7>. Using the right-hand side of *axiom Q4*,
Expression (4) is the same as

if **isemptyqueue** (y)
then
    1
else
    front (y)   *(5)*

The required result is therefore obtained by

    front (y).   *(6)*

Writing *Expression (6)* in the form of the left-hand side of *axiom Q4* gives

    front (**addtoqueue** (7, **createqueue**))   *(7)*

and the right-hand side allows *Expression (7)* to be written as

if **isemptyqueue** (**createqueue**)
then
    7
else
    front (**createqueue**)   *(8)*

In *Expression (8)* the *then* is the relevant part, giving the final result of 7. Notice that the recursion is simpler here than the case with **deletefromqueue**.

5.3 None! The fact that our implementation includes a dummy record means that we do not have to take special action when adding the first item to a queue.

5.4 The main change is that special action would have to be taken when adding a first item to a queue. Also, a different representation for an empty queue would be needed. Typically we would choose to represent an empty queue by storing the **nil** pointer in the *Head* field.

5.5 (i) The structure is empty when the *Head* and *Tail* pointers point to the same record.

(ii) The structure is full when the *Head* pointer is one position ahead of the *Tail* pointer.

5.6 The heading of the print routine would be placed in the **interface** part of the unit, and the body (together with the name of the routine) would be placed in the **implementation** part of the unit. The unit would then have to be recompiled, as would every application program that uses the unit.

5.7 **procedure** *PrintQueue (Q: Queue);*
       var
          Current: Cursor;
       begin
          if not *IsEmptyQueue (Q)*
          then
             begin
                Current := Q.Head;
                repeat
                   GetItem(Q.S, Current, Item);
                   writeln(Q.Item);
                   Current := MoveCursor (Current)
                until *Current = Q.Tail*
             end
       end *{PrintQueue};*

Chapter 6

6.1 **createdeque** is equivalent to **createqueue**
       isemptydeque is equivalent to **isemptyqueue**
       produceleft is equivalent to **front**
       addright is equivalent to **addtoqueue**
       deleteleft is equivalent to **deletefromqueue**

A queue does not support operations equivalent to **produceright, addleft,** or **deleteleft.**

6.2 *NAME*
          deque(item)

       SETS
          D   the set of deques
          I   the set of items
          B   {*true, false*}
          M   {*error — empty deque*}

6.3   *SYNTAX*
             createdeque: $\rightarrow D$
             isemptydeque: $D \rightarrow B$
             produceleft: $D \rightarrow I \cup M$
             produceright: $D \rightarrow I \cup M$
             addleft: $I \times D \rightarrow D$
             addright: $I \times D \rightarrow D$
             deleteleft: $D \rightarrow D \cup M$
             deleteright: $D \rightarrow D \cup M$

6.4   The underlying model used in the following specification is that of a list. Hence, $\forall$ i $\in$ I; d $\in$ D; b $\in$ B; s $\in$ D $\cup$ M; and r $\in$ I $\cup$ E:

```
pre-createqueue ( ) ::= true
post-createqueue (d) ::= (d = createlist )
(or, post-createqueue (d) ::= isemptylist (d))
```

```
pre-isemptydeque (d) ::= true
post-isemptydeque (d; b) ::= (b = isemptylist (d))
```

```
pre-produceleft (d) ::= true
post-produceleft (d; r) ::= if  d = createlist
                           then  r = error — empty deque
                           else  r = first (d)
```

```
pre-addleft (i, d) ::= true
post-addleft (i, d; s) ::= (s = concatenate (make (i), d))
```

```
pre-deleteleft (d) ::= true
```

```
post-deleteleft (d; r) ::=  if  d = createlist
                           then  s = error — empty deque
                           else  s = trailer (d)
```

The operations **produceright, addright,** and **deleteright** are similar to **produceleft , addleft,** and **deleteleft** respectively.

6.5   As was mentioned in *Chapter 3* there is often an awkwardness when adding the first item to a data structure. The difficulty can be overcome either by additional programming within the implementation of the operations, or by including a dummy item in the sequence. Either method is acceptable but, in contrast to what we did earlier, we shall avoid the use of a dummy item in our solution. Here is a deque with three items:

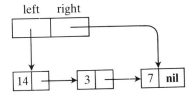

An empty deque:

left	right
nil	**nil**

A deque with one item only:

The last item has its *Next* field set to **nil** to indicate the end of the sequence (going from left to right, following the arrows).

6.6 The left pointer is updated. (The *Next* field of the new deque item points to the previous left item.)

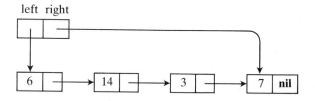

6.7 The only link to be updated is the one pointing to the left item.

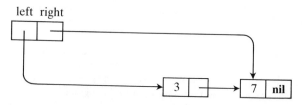

6.8 Two links have been updated. The right pointer now points to the previously penultimate item which, in turn, has had its *Next* field set to **nil**.

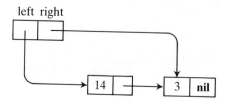

6.9 **type**
  Link = ↑*DequeRecord;*
  DequeRecord = **record**
        Item: ItemType;
        Next: Link
      end;
  Deque = **record**
       Left, Right: Link
      end;

6.10 The form of the procedure depends on whether use is being made of a dummy record in the representation. Here it is not. The required procedure is given below.

```
procedure CreateDeque (var  D: Deque);
begin
  with D do
  begin
    Left := nil;
    Right := nil
  end
end {CreateDeque};
```

6.11 **function** *IsEmptyDeque (D: Deque): Boolean;*
  begin
   *IsEmptyDeque := (D.Left = **nil**) and (D.Right = **nil**)*
  end *{IsEmptyDeque};*

Such a statement as:

  IsEmptyDeque := (Left = Right)

will fail to distinguish between an empty deque and a deque containing a single item.

6.12 **procedure** *ProduceLeft (D: Deque;* **var** *Item: ItemType);*
  begin
   if *IsEmptyDeque(D)*
   then
    writeln('Error — empty deque')
   else
    Item := D.Left↑. Item
  end *{ProduceLeft};*

6.13 **procedure** *AddLeft (Item: ItemType;* **var** *D: Deque);*
    var
      Temp: Link;
    begin
      new(Temp);
      Temp↑. Item := Item
      if *IsEmptyDeque(D)*
      then *{add to empty deque}*
        begin
          Temp↑. Next := **nil;**
          with *D* **do**
            begin
              Left := Temp;
              Right := Temp
            end
        end
      else *{items already present}*
        begin
          Temp↑. Next := D.Left;
          D.left := Temp
        end
    end *{AddLeft};*

6.14 **procedure** *DeleteLeft (*var* D: Deque);*
    var
      Temp: Link;
    begin
      Temp := D.Left;
      D.Left := D.Left↑. Next;
      Dispose (Temp)
    end *{DeleteLeft};*

6.15 **procedure** *DeleteRight (*var* D: Deque);*
    var
      CurrentItem, PreviousItem: Link;
    begin
      if *IsEmptyDeque(D)*
      then
        writeln('Error — empty deque')
      else
        if *(D.Left = D.Right)* **and** *(D.Left <>* nil *)*
        then *{only one item in deque}*
          begin
            D. Left := **nil;**
            D. Right := **nil**
          end

```
      else {more than one item in deque}
        begin
          CurrentItem := D.Left;
          PreviousItem := nil;
          while CurrentItem.Next <> nil do
            begin
              PreviousItem := CurrentItem;
              CurrentItem := CurrentItem↑.Next
            end;
          PreviousItem↑.Next := nil;
          D.Right := PreviousItem;
          Dispose(CurrentItem)
        end
  end {DeleteRight};
```

6.16 Here is the UCSD Pascal **interface** part:

```
unit DequeOps;
interface
type
  ItemType = Integer;
  Link = ↑DequeRecord;
  DequeRecord = record
                  Item: ItemType;
                  Next: Link
                end;
  Deque =  record
             Left, Right: Link
           end;

procedure CreateDeque (var D: Deque);
function IsEmptyDeque (D: Deque): Boolean;
procedure AddLeft (Item: ItemType;  var D: Deque);
procedure AddRight (Item: ItemType;  var D: Deque);
procedure ProduceLeft (D: Deque;  var Item: ItemType);
procedure ProduceRight (D: Deque;  var Item: ItemType);
procedure DeleteLeft (var D: Deque);
procedure DeleteRight (var D: Deque);

{end of interface part}
```

The equivalent MODULA-2 definition module is given next.

```
DEFINITION MODULE DequeOps;
  TYPE ItemType = INTEGER;
  TYPE Link = POINTER TO DequeRecord;
  TYPE DequeRecord = RECORD
                          Item: ItemType;
                          Next: Link;
                        END;
  TYPE Deque =   RECORD
                    Left, Right: Link;
                  END;

  PROCEDURE CreateDeque ( VAR d: Deque);
  PROCEDURE IsEmptyDeque (d: Deque): BOOLEAN;
  PROCEDURE AddLeft (i: ItemType; VAR d: Deque);
  PROCEDURE AddRight (i: ItemType; VAR d: Deque);
  PROCEDURE ProduceLeft (d: Deque; VAR i: ItemType);
  PROCEDURE ProduceRight (d: Deque; VAR i: ItemType);
  PROCEDURE DeleteLeft ( VAR d: Deque);
  PROCEDURE DeleteRight ( VAR d: Deque);
END DequeOps.
```

6.18
```
program TestDequeOps;
uses DequeOps;        {or whatever is required by your system}
var
  Item: ItemType;
  FirstDeque, SecondDeque: Deque;
  i: Integer;
begin
  CreateDeque(FirstDeque);
  CreateDeque(SecondDeque);
  for i:=1 to 4 do
    begin
      writeln('Enter item ', i:2, ': ');
      readln (Item);
      AddLeft (Item, FirstDeque)
    end;
  for i:=1 to 4 do
    begin
      ProduceRight (FirstDeque, Item);
      DeleteRight (FirstDeque);
      AddRight (Item, SecondDeque)
    end;
  for i:=1 to 4 do
    begin
      ProduceLeft (SecondDeque, Item);
      writeln (Item);
      DeleteLeft (SecondDeque)
    end
end {TestDequeOps}.
```

6.19 procedure *PrintDeque (D: Deque);*
  var
    CurrentItem: Link;
  begin
    CurrentItem := D.Left;
    while *CurrentItem <>* **nil do**
      begin
        writeln (CurrentItem ↑.Item);
        CurrentItem := CurrentItem↑.Next
      end
  end *{PrintDeque}*;

6.20 A system of two-way links would improve the representation. That is, each item should be linked to both its neighbours, on the left as well as on the right. This requires that each item be accompanied by two link fields.

6.21 (ii) No changes should be necessary to any application program using the abstract data type deque.

Chapter 7

7.1 (i) There are 9 leaf nodes: North, Midlands, South West, South East, Highlands, Borders, Northern Ireland, North, and South.

(ii)   The root node has 4 children: England, Scotland, Northern Ireland, Wales.

(iii)   The parent node of Midlands is England.

(iv)   There are no child nodes of Midlands — it is a leaf node!

(v)   There is one child node of Lowlands: Borders.

(vi)   This is the root node — it does not have a parent!

7.2 (i)   The root node contains the item England.

(ii)   There are four leaf nodes: North, Midlands, South West, South East.

(iii)   There are three child nodes of the root node: North, Midlands, South.

(iv)   There are 5 possible subtrees. Here are their root items: North, Midlands, South, South West, South East.

7.3 There are three binary trees shown. The left hand tree is the only one which is not binary, having more than two children from one of its nodes (England).

7.4

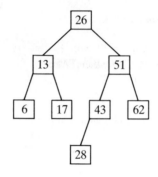

7.5 **data** yields the result *Tom*. **left** and **right** yield the following trees:

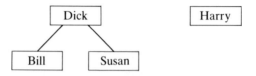

Note that the result of applying **right** is a tree with only one item, and is not simply the item itself.

7.6

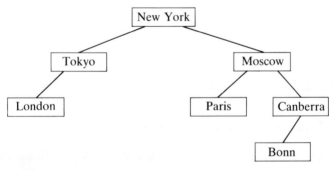

7.7 **createtree:** → T *(BST1)*
   isemptytree: T → B *(BST2)*
   data: T → I ∪ M *(BST3)*
   left: T → T *(BST4)*
   right: T → T *(BST5)*
   maketree: T × I × T → T *(BST6)*

7.8 $\forall i \in I;\ l, r \in T:$
isemptytree (**createtree**) = *true*
isemptytree (**maketree** (l, i, r)) = *false*
data (**createtree**) = *empty tree*
data (**maketree** (l, i, r)) = i
left (**createtree**) = **createtree**
left (**maketree** (l, i, r)) = l
right (**createtree**) = **createtree**
right (**maketree** (l, i, r)) = r

7.9 $\forall i \in I,\ l, r \in T:$
number_in (**createtree**) = 0
number_in (**maketree** (l, i, r)) = 1 + **number_in** (l) + **number_in** (r)

7.10 *NAME*
BSTree (item)

SETS
Bst   the set of binary search trees
I   the set of items
B   {*true, false*}
M   {*empty tree*}

SYNTAX
createtree: $\rightarrow$ Bst
isemptytree: Bst $\rightarrow$ B
data: Bst $\rightarrow$ I $\cup$ M
left: Bst $\rightarrow$ Bst
right: Bst $\rightarrow$ Bst
maketree: Bst $\times$ I $\times$ Bst $\rightarrow$ Bst
isin: I $\times$ Bst $\rightarrow$ B
insert: I $\times$ Bst $\rightarrow$ Bst

7.11 Since **left** applied to an empty BSTree yields an empty BSTree we have:

left (**createtree**) = **createtree**.   *(T3)*

left applied to a non-empty BSTree yields the left subtree, so:

left (**maketree** (l, i, r)) = l.   *(T4)*

7.12 **right** (**createtree**) = **createtree**   *(T5)*
right (**maketree** (l, i, r)) = r.   *(T6)*

7.13 Since **data** applied to an empty BSTree yields an error value we have:

> **data** (**createtree**) = *empty tree.*   (T7)

data applied to a non-empty BSTree yields the root item, so:

> **data** (**maketree** (l, i, r)) = i.   (T8)

7.14 This solution is complete, in that all necessary steps are given, but the reasoning behind each step has been abbreviated. The question asks for the value of the expression **isin** (New York, t); the evaluation proceeds as shown below (you will have to draw out the subtrees for yourself).

isin (New York, t)
   = **isin** (New York, **maketree** (l, London, r))   {*by definition of maketree*}
   = **isin** (New York, r)   {*by second axiom since New York > London*}
   = **isin** (New York, **maketree** (l', Paris, r'))
   = **isin** (New York, l')   {*by second axiom since New York < Paris*}
   = **isin** (New York, **maketree** (l", Moscow, r"))   {*l" and r" are empty trees*}
   = **isin** (New York, r")
   = **isin** (New York, **createtree**)   {*since r" is empty*}
   = *false*   {*by first **isin** axiom*}

7.15 The expression **insert**(e, l) results in a BSTree which consists of the BSTree *l* with a new node containing the item *e* added to it. The given expression constructs a new BSTree whose left subtree is the BSTree which results from **insert** (e, l), whose right subtree is *r* and whose root item is *i*.

7.16

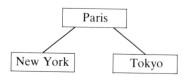

7.17 The first insert axiom can be used to add the first item, *Oslo:*

> **insert** (Oslo, **createtree**) = **maketree** (**createtree** , Oslo, **createtree**)

which is:

> | Oslo |

If we denote this tree by *t,* then inserting the item *Beijing* is achieved with

> **insert** (Beijing, t)

which is the same as

insert (Beijing, **maketree** (**createtree** , Oslo, **createtree**)).

The second axiom can now be employed to give

maketree (**createtree** , Oslo, **insert** (Beijing, **createtree**))

(since *Beijing* > *Oslo*). Now employ the first axiom to get

maketree (**createtree** , Oslo, **maketree** (**createtree** , Beijing, **createtree**))

which is the tree:

7.18 (a) Print out the left subtree (consisting of the single node containing *Moscow*);
    (b) Print out *Paris* (the root node data value);
    (c) Print out the right sub-tree (consisting of the single node containing *Tokyo*).

7.19 (a) The data are printed out in the order: 11, 14, 21, 23, 8, 2, 7, 9, 6, 26. Note that the data are *not* printed out in numerical order because the tree is not a binary search tree. This exercise shows that the inorder traversal can be applied to any binary tree, and that the data will be printed out in the order in which it is stored in the tree.

7.20 The data is printed out in the following order:
    London, Canberra, Bonn, Paris, Moscow, Tokyo.

7.21 if the tree is not empty
    then
        print out the left subtree
        print out the right subtree
        print out the root node data value
    if end

7.22 For **preorder** the two axioms are

preorder (**createtree**) = **createqueue**   *(T15)*
preorder (**maketree** (l, i, r)) =
    appendqueue (**appendqueue** (**addtoqueue** (i, **createqueue**),
                preorder (l)),
                preorder (r))   *(T16)*

and for **postorder** the two axioms are

> postorder (createtree) = createqueue   *(T17)*
> postorder (maketree (l, i, r)) =
>   appendqueue (appendqueue (postorder (l),
>   postorder (r))
>   addtoqueue (i, createqueue))   *(T18)*

Chapter 8

8.1 $\forall$ e, n $\in$ N; t $\in$ T; and d $\in$ D:

remove (e, **createdirectory**) = *employee not in directory*

remove (e, **makedirectory** (n, t, d)) =
   (e = n : d   {*the subdirectory with the first item removed*}
   | e < n : *employee not in directory*
   | e > n : **makedirectory** (n, t, **remove** (e,d)))

8.2 **procedure** *Retrieve (n: Name; d: Directory;* **var** *t: PhoneNo);*
var *Current: Link;*
     Found: Boolean;
begin
   Found := false;
   Current := d$\uparrow$. Next;
   while *(Current <>* **nil** *)* **and not** *Found* **do**
      if *n = Current$\uparrow$. EmployeeName*
      then
         begin
            t := Current$\uparrow$. Telephone;
            Found := true
         end
      else
         if *n < Current$\uparrow$. EmployeeName*
         then
            Current := **nil;**   {*to force searching to stop*}
   if not *Found*
   then
      Writeln('Employee not in directory.')
end *{Retrieve};*

8.3   **program** *ChangeIt;*
      uses *DirectoryUnit;*
      var *n: Name;*
         t: PhoneNo;
      begin
        Writeln('Enter Employee name: ');
        Read(n);
        Writeln ('Enter new phone number: ');
        Read(t);
        if *IsInDirectory (n, d)*
        then
          begin
            Remove(n, d);
            Add (n, t, d)
          end
      end *ChangeIt.*

8.4 Calling the two lists the *left list* and the *right list,* we can designate the current record to be *either* the record at the tail of the left list *or* the record at the front of the right list. We have chosen the latter.

post-**createfile** (f) ::= f = < **createlist, createlist** >

post-**isemptyfile** (< l, r >; b) ::= b = (r = **createlist**)

post-**insert** (< l, r >, i; f) ::= f = < **concatenate** (l, **first** (r)),
                                concatenate (**make** (i), **trailer** (r)) >

post-**forward** (< l, r >; f) ::=
                if r = **createlist**
                then
                  f = *error*
                else
                  if **trailer** (r) = **createlist**
                  then
                    f = < l, r >
                  else
                    f = < **concatenate** (l, **first** (r)), **trailer** (r) >

post-**backward** (< l, r >; f) ::=
                if r = **createlist**
                then
                  f = *error*
                else
                  if l = **createlist**
                  then
                    f = < l, r >
                  else
                    f = < **header**(l), **concatenate** (**last** (l), r) >

post-**replace** (< l, r >, i; f) ::= *if* r = **createlist**
                                          then
                                             f = *error*
                                          else
                                             f = <l, **concatenate** (**make** (i), **trailer** (r)) >

8.5   **PROCEDURE** *IsEmptyFile(f: File): BOOLEAN;*
          BEGIN
             RETURN *(f=* **NIL***)*
          END *IsEmptyFile;*

          PROCEDURE *Delete(VAR f: File);*
          VAR *Temp: File;*
          BEGIN
            IF *f =* **NIL**
            THEN
                WriteString('Empty File.'); WriteLn
            ELSE
               IF *f* ↑*.Left* = **NIL**
               THEN
                  IF *f* ↑*.Right* = **NIL**
                  THEN
                     f:= **NIL**
                  ELSE
                     f:=f ↑*.Right*
                  END
               ELSE
                  f↑*.Left*↑*.Right := f*↑*.Right;*
                  f↑*.Right*↑*.Left := f*↑*.Left;*
                  Temp := f;
                  f:=f ↑*.Left;*
                  Deallocate(Temp)
               END
            END
          END *Delete;*

          PROCEDURE *Backward (VAR f: File);*
          BEGIN
            IF *f =* **NIL**
            THEN
                WriteString('Empty File.'); WriteLn
            ELSE
               IF *f* ↑*.Left* <>**NIL**
               THEN
                  f:=f ↑*.Left*
               END
            END
          END *Backward;*

8.6 $\forall$ i, e $\in$ I (the set of items); s, t $\in$ S (the set of all simple sets):

in (i, **createset**) = *false*

in (i, **makeset** (e, s)) = *if* i = e
          then
             true
          else
             in (i, s)

delete (i, **createset**) = **createset**

delete (i, **add** (e, s)) = *if* i = e
            then
               delete (i, s)
            else
               add (**delete** (i, s), e)

Chapter 9

9.1

Operation	Object	Type	Changed?
Delivery	Stock	Quantities	Yes
	Delivery note	Quantities/items	No
Order	Stock	Quantities	Yes
	Invoice	Quantities/items	No
Print stock value	Stock	Quantities	No
	Prices	Money	No
Print stock holding	Stock	Quantities	No
Read delivery note	Delivery note	Quantities/items	Yes
Read invoice	Invoice	Quantities/items	Yes
Initialize stock	Stock	Quantities/items	Yes

9.2 In Ada the storage space for an instance is allocated at compile time (as the result of a variable declaration). At the same time an identifier is associated with the instance. The initialization of private data occurs at run-time when a 'create' procedure is called (e.g. **CreateStack**). In an object oriented-language creation of an object, its identification, and initialization all occur at run time when a new message is sent to a class.

Index

abstract model 39
abstraction 14
Ada 54, 84
algebraic approaches *see* axiomatic
Apple Macintosh computer 215
application independent 66
applications programmer 64
array representation of stack 56
attributes 198, 202
auxiliary set 18
aximomatic approach 15, 25, 158, 199
axioms 10, 26

binary tree 150, 154
binary search tree (BSTree) 152, 160
   implementation (MODULA-2) 175
branch 149

canonical form 37
carrier set 18
Cartesian product 20
child node 150
circular structure 132
class 218
completeness 34
composition 34
concatenate 183
constraint 23, 77
constraint violation 78, 109
constructive approach 15, 38, 191
convert 183
correctness 12
cursor-based implementation 54

data administrator 198
database specification 199
   implementation 202
decomposition operators 157
definition module 105
deque 142
   − linked representation 144
   pointer based implementation 147
desktop 214
directory 180
   specification 182
   implementation 186

dot-notation 96, 104
double ended queue *see* deque
dummy record 125

encapsulation 77, 84
error situations 22
exception handling 109
exported 86
expressions 15
external objects 104

file processing 189
   specification 191
   implementation 194
formal language 12
formal specification 12

generic abstact data type 77, 218
generic concept 215
generic instantiation 98
generic package 98
generic parameter 99
generic specification 17, 101, 158, 191
guarded operation 78

head 117
hidden operation 51, 161, 183, 199

implementation part 70
implementation 12, 53
implementation module 105
implementors 64
import 86
import list 106
information hiding 54, 63, 84, 215
inherent constraint 23
inheritance 220
input/output routines 85
instance methods 218
instance variable 218
interface part 72
invariant assertion 36, 42, 50, 119

leaf node 149
linear structure 39
linked list, two-way 194